TROPICAL SOIL BIOLOGY AND FERTILITY

A Handbook of Methods

Second Edition

Edited by

J.M. Anderson
Department of Biological Science
Exeter University
Exeter EX4 4PS
UK

and

J.S.I. Ingram
IGBP-Global Change and Terrestrial Ecosystems
Focus 3 Office
Department of Plant Sciences
University of Oxford
Oxford OX1 3RB
UK

CAB INTERNATIONAL

CABI *Publishing*
A division of CAB INTERNATIONAL

CABI *Publishing*
Wallingford
Oxon OX10 8DE
UK

Tel: +44 (0)1491 832111
Fax: +44 (0)1491 833508
Email: cabi@cabi.org

CABI *Publishing*
10 E 40th Street, Suite 3203
New York, NY 10016
USA

Tel: +1 212 481 7018
Fax: +1 212 686 7993
Email: cabi-nao@cabi.org

A catalogue record for this book is available from the British library

ISBN 0 85198 821 0

First printed 1993
Reprinted 1996, 1998

Printed and bound in the UK by Information Press Ltd, Eynsham

Table of Contents

Preface to the Second Edition

During the time since publication of the first edition of this Handbook in 1989 the methods have been tested in a wide range of field and laboratory environments. This both intensive and extensive testing has lead to considerable debate and proposals for revision; we hope this second edition adequately reflects the substantial collective thought, experience and input of many scientists.

Overall the Handbook now comprises seven chapters and 15 appendices.

A completely revised Chapter 1 now complements the TSBF philosophy with a discussion of an example TSBF-type experiment. This sets the recommended site characterisation and methodology in better context. The first edition main text has been substantially re-structured to offer a more logical and sequential approach to both field and laboratory procedures. A condensed Chapter 2 discusses site characterisation, while Chapters 3 - 7 cover field measurements, field sampling, laboratory practice and quality control, chemical analyses and soil physical analyses, respectively. Additional material on sample preparation and the new chapter on laboratory practice and quality control provide a more rounded laboratory compendium. Subsection introductions have been developed and, where appropriate, cited and "suggested further reading" references updated. Several new diagrams have been included and a standardised, and clearer format used throughout.

Many users found the appendices of great value. Most appendices from the first edition have therefore been revised, and several (socioeconomic considerations, soil phosphorous, near-surface microclimate and suggested equipment and reagents) have been completely re-written. Several new appendices have been added (mycorrhizas, roots, most probable number analysis for rhizobia, isotope studies and a description of the CENTURY model).

We believe this second edition now forms a substantially clearer and more complete reference work for ecosystem studies, being equally applicable to both tropical and temperate systems.

Jonathan Anderson and John Ingram
October 1992

Acknowledgements

The Tropical Soil Biology and Fertility Programme
was established as one of the four programmes of the
International Union of Biological Sciences / UNESCO
Man and the Biosphere Programme

"The Decade of the Tropics".

Financial assistance from the Rockefeller Foundation and the International Soil Science Society for the preparation of this Handbook is gratefully acknowledged.

The Editors gratefully acknowledge the editorial assistance of Dr Mary Scholes and Dr Robert Scholes, and the assistance of Alice N'dungu in manuscript preparation.

The methods given in this Handbook are endorsed by
The International Society of Soil Science.

Principal Contributors

I Alexander, *University of Aberdeen, UK*
J M Anderson, *Exeter University, UK*
K F Baker, *Woodside, Pelican Road, Pamber Heath, Hants, UK*
E Barrios, *Instituto Venezolano de Investigaciones Cientificas, Caracas, Venezuela*
L Bohren, *Colorado State University, Fort Collins, USA*
G D Bowen, *CSIRO Division of Soils, Glen Osmond, Australia*
S Brown, *US Environmental Protection Agency, Corvallis, USA*
E Cuevas, *Instituto Venezolano de Investigaciones Cientificas, Caracas, Venezuela*
J M Dangerfield, *University of Botswana, Gaberone, Botswana*
T Darnhofer, *United Nations Environment Programme, Nairobi, Kenya*
Y Dommergue, *ORSTOM, Nojent sur Marne, France*
E T Elliott, *Colorado State University, Fort Collins, USA*
K Giller, *Wye College, University of London, UK*
M Hornung, *Institute of Terrestrial Ecology, Merlewood, UK*
J S I Ingram, *IGBP-Global Change and Terrestrial Ecosystems, Oxford, UK*
J Ladd, *CSIRO Division of Soils, Glen Osmond, Australia*
P Lavelle, *ORSTOM, Bondy, France*
R Merckx, *Katholieke University, Leuven, Belgium*
R J Myers, *International Board for Soil Research and Management, Bangkok, Thailand*
L Nelson, *North Carolina State University, USA*
M van Noordwijk, *International Center for Research in Agroforestry, Bogor, Indonesia*
C A Palm, *Tropical Soil Biology and Fertility Programme, Nairobi, Kenya*
M B Peoples, *CSIRO Division of Plant Industry, Canberra, Australia*
R J Raison, *CSIRO Division of Forestry, Canberra, Australia*
D J Read, *Sheffield University, UK*
P Robertson, *WK Kellogg Biological Station, Hickory Corners, USA*
A P Rowland, *Institute of Terrestrial Ecology, Merlewood, UK*
P A Sanchez, *International Center for Research in Agroforestry, Nairobi, Kenya*
M C Scholes, *Witwatersrand University, Johannesburg, Republic of South Africa*
R J Scholes, *CSIR Forest Science and Technology, Pretoria, Republic of South Africa*
P D Seward, *Tropical Soil Biology and Fertility Programme, Nairobi, Kenya*
A R A Stapleton, *Exeter University, UK*
J L Stewart, *Oxford Forestry Institute, UK*
M J Swift, *Tropical Soil Biology and Fertility Programme, Nairobi, Kenya*
H Tiessen, *University of Saskatchewan, Saskatoon, Canada*
P M Vitousek, *Stanford University, USA*
B H Walker, *CSIRO Division of Wildlife and Ecology, Canberra, Australia*
P L Woomer, *Tropical Soil Biology and Fertility Programme, Nairobi, Kenya*
A Young, *University of East Anglia, Norwich, UK*

Chapter 1 THE TROPICAL SOIL BIOLOGY AND FERTILITY PROGRAMME, TSBF

1.1 THE TSBF RATIONALE, PHILOSOPHY AND BACKGROUND

Recent years have seen a dramatic increase in the *per capita* food production in much of the tropics. This improvement is largely based on the introduction of new crop varieties into farming programmes on fertile soils with good supplies of water, fertiliser and pesticide. In large parts of Africa, however, and in many other less fertile parts of the tropics, the production trend is the opposite and *per capita* production of food has actually been declining for twenty years.

The further spread of high-input farming systems is undoubtedly one answer to this situation, and one which is being pursued vigorously by the development of new and appropriate technologies. It is apparent however that the economic cost of such agriculture is often unacceptably high. These farming systems are also commonly of low efficiency in terms of resource use and may be accompanied by rapid environmental degradation. At the same time the more environmentally conservative traditional forms of agriculture practised in much of the tropics are no longer sustainable because of increased population densities and pressures on land. The last two decades have therefore seen great interest in the development of farming systems characterised by a relatively inexpensive level of input, a high efficiency of internal resource use and hence more sustainable production in both economic and ecological terms.

Sustainable use of the soil resource is a primary goal of all such farming systems and it is to this target that the Tropical Soil Biology and Fertility Programme (TSBF) is directed. The Programme's stated general objective is "to determine the management options for improving tropical soil fertility through biological processes" (Swift, 1984). TSBF differs from, and complements, other research programmes in concentrating on research into the biological resources and processes of soil; it draws on expertise from both agricultural and ecological research.

In natural ecosystems productivity is sustained by the tight integration of the vegetative system with the biological system of soil in relation to key processes such as nutrient cycling and the formation and breakdown of soil organic matter. These crucially important biological processes of soil are still poorly understood by ecologists, even in natural ecosystems, and are rarely investigated by agriculturalists. One of the reasons for this is the success of high-input farming, which effectively bypasses soil biological processes through its use of fertilisers, pesticides and the mechanised preparation of soil.

This success leaves little apparent reason why soil biological processes should be taken seriously. The focus on sustainable low-input agriculture described above does however provide such a rationale. Furthermore, it is noticeable that the adoption of minimum tillage systems in temperate regions is refocusing attention on soil ecology. Such systems can provide economically viable options which may be particularly applicable to the sensitive soils of certain parts of the tropics. Moreover, much of our current knowledge of these soil biological processes has been achieved through pure research just as the understanding of the consequences of perturbing nutrient cycles has largely come from manipulative studies of natural ecosystems. It is therefore necessary to retain and develop an interface between natural and agroecosystems as a context for integrating and building on the knowledge of ecologists and agricultural scientists. It is evident that there are fundamental as well as practical scientific problems to be solved; these problems are not unique to the tropics but it is the particular problems of the tropics that demand their solution.

TSBF was initiated in 1984 by the International Union of Biological Sciences and the UNESCO *Man and Biosphere* Programme. The underlying aim is to stimulate research in the poorly-understood topic of the role of biological processes in the maintenance of soil fertility. It is one component of the collaborative research programme entitled "The Decade of the Tropics", the objective of which is to "*increase our knowledge and understanding of the biology of the tropics from the point of view of the various biological subdisciplines*" (Solbrig and Golley, 1983).

The TSBF strategy includes focus on the processes of carbon, nitrogen and water transfers through the soil-plant system, and the key factors which regulate them. The practical targets are to synchronise plant demand with nutrient and water availability, and to manipulate and conserve soil organic matter pools. The tools with which this can be achieved are the management of the quantity, quality and timing of organic inputs, and the manipulation of soil organisms. To understand better the regulatory processes which underpin natural systems, and how they differ for managed systems, a key tactic has been the comparison of managed systems with the natural ecosystems from which they were derived.

Further details on TSBF's background can be found in Swift (1984, 1987), Ingram and Swift (1989) and Woomer and Ingram (1990); a comprehensive synthesis of TSBF and associated research approach and findings is presented in Woomer and Swift (1994).

References

Ingram, J.S.I. and Swift, M.J. (eds.) (1989) Tropical Soil Biology and Fertility Programme: report of the Fourth International Workshop, Harare. *Biology International* Special Issue **20**, 1-64.

Solbrig, O.T. and Golley, F.B. (1983) A decade of the tropics. *Biology International* Special Issue **2**, 1-15.

Swift, M.J. (ed.) (1984) Soil biological processes and tropical soil fertility (TSBF): planning for research *Biology International* Special Issue **9**, 1-24.

Swift, M.J. (ed.) (1987) Tropical Soil Biology and Fertility: Inter-regional research planning workshop, Yurimaguas. *Biology International* Special Issue **13**, 1-68.

Woomer, P.L. and Ingram, J.S.I. (eds.) (1990) *The Biology and Fertility of Tropical Soils.* TSBF Report: 1990. TSBF, Nairobi.

Woomer, P.L. and Swift, M.J. (eds.) (1994) *The Biological Management of Tropical Soil Fertility.* John Wiley and Sons, Chichester.

1.2 TSBF RESEARCH CONCEPTS

Over the past few years, research has been conducted within TSBF to address two major Principles that consider the relationships between soil biological processes and soil fertility.

The first is termed "Synchrony", and recognises that the availability of nutrients to plants is controlled by a complex set of interactions. Nevertheless, the release of nutrients from decomposing organic residues can be synchronised with plant growth demand. A dynamic equilibrium exists between nutrient supply from mineralisation (or addition of inorganic fertiliser) and its control by immobilisation on organic materials and losses through volatilisation, denitrification and leaching. Biological processes regulate this equilibrium, and, as they are open to potential manipulation, the additional competing process of plant uptake can be optimised. It is hypothesised that the potential for such manipulation will be greatest when there is a diversity of resources available, and when management options are relatively flexible. The main way to promote synchrony is with the management of the timing, placement and quality of organic residues.

The second major Principle concerns soil organic matter. Soil organic matter plays key roles in crop sustainability, primarily through its interactions with soil chemical and physical properties in relation to nutrient release, cation retention and soil structure. The value of soil organic matter (as distinct from the value of organic inputs described in the "Synchrony" Principle) is well recognised, but little is known about the processes that contribute to its three key roles. This is in sharp contrast with the well understood processes underlying the use of chemical fertilisers. As low-input systems regain importance in the tropics it is essential to improve understanding of the functioning of soil organic matter.

Soil organic matter can differ in both quantity and quality (composition) resulting in different patterns of nutrient release and availability. The quality of organic inputs may affect the composition, and the long- and short-term nutrient release rates from soil organic matter. The benefits of soil organic matter on soil properties associated with healthy plant growth are well demonstrated; the processes which lead to the accumulation of organic matter in soils are however extremely complex, being associated with overall environmental conditions. To assist in rationalising the complex processes inherent in the maintenance of soil fertility, simulation models are often used. The TSBF Programme has adopted the CENTURY model to simulate plant production, nutrient cycling and soil organic matter dynamics for TSBF sites. A central feature of CENTURY is the division of soil organic carbon into functional pools based on the residence time in soils. A description of the CENTURY model, and its application in TSBF studies, is given in Appendix L ("Use of the CENTURY plant-soil environmental simulation model").

Both these major Principles incorporate consideration of the themes of soil water and soil fauna. All soil biological and chemical activities are dependent on an adequate level of soil water. Fluctuating moisture conditions are a feature of most tropical areas, and both an excess or a shortage of water in the soil can limit decomposition, nutrient release, nutrient uptake by plants and plant growth. The duration for which water is available within the tolerance range of a particular process therefore has an overall controlling influence on the degree to which that process can operate.

Soil fauna, particularly earthworms and termites, are important as regulators of decomposition, nutrient cycling and soil organic matter dynamics; they directly and indirectly effect soil structure and hence aeration, water infiltration and water holding capacity. The response of soil fauna to changes in land use and cultivation practices can therefore impart profound effects on soil fertility through their feeding and burrowing activities. To help understand their roles in nutrient regulation and soil organic matter turnover, soil fauna can be classified into ecological groups based on their feeding habits and habitats.

The TSBF experimental approach has evolved from the recognition that the ability of natural plant communities to conserve nutrients and moisture, through the effects of canopy, litter and soil organic matter, serve as an effective model for the development of more sustainable agroecosystems. Soil fertility is therefore seen as the integration of plant nutrient demand, decomposition processes, soil fauna activities and their interaction with soil chemical and physical properties. These processes are inextricably linked, and cultivation practices such as green manuring, mulching or crop residue incorporation have a multiplicity of indirect effects on soil biological processes. The suite of biological, physical and chemical variables used in TSBF studies, provide a basis for interpreting the complex direct and indirect effects of treatments on soil processes and conditions.

1.3 TSBF EXAMPLE "SYNCHRONY" EXPERIMENT

As discussed above, the potential to synchronise nutrient availability with plant demand through the judicious use of crop residues offers a means of improving crop productivity while reducing nutrient losses. In many agroecosystems, crop residues are however used in a number of ways, each of comparative value, as e.g. animal feed, fuel or organic additions to subsequent crops. Furthermore, many farming systems have access to only limited quantities of nutrient-rich (high quality) plant materials or animal wastes. Synchrony experimentation therefore seeks to maximise nutrient use-efficiency through the combinations of inputs of contrasting chemical composition. Although this example experiment considers only two materials of contrasting quality, several of the central features of synchrony-type field experimentation apply:

- relatively few treatments are examined, but each in great detail, in order to identify the pools and fluxes of the nutrient that limits plant productivity.

- equal amounts of a potentially limiting plant nutrient are applied in different forms and combinations to all treatments (except the complete control); consequently, differences observed between treatments result from the use-efficiency of the various resources rather than the amounts applied.

- frequent plant sampling is required in order to estimate plant nutrient uptake and demand.

- measurement of the effects of residue addition on several soil properties in addition to mineralisation of the nutrient limiting plant productivity; i.e. soil faunal dynamics, organic matter fractionation, soil water content and water holding capacity.

1.3.1 General hypothesis

The release of nutrients from above-ground inputs and decomposing roots can be synchronised with plant growth demands.

1.3.2 Working hypotheses

- mixtures of applied organic residues of contrasting chemical properties result in improved plant productivity through controlling the pattern of nutrient release.

- residue application results in short-term immobilisation of plant nutrients within the soil microbial biomass that becomes available to the crop later in the growing season when nutrient demands are greater.

1.3.3 Experimental treatments

This example experiment uses locally available organic materials. The amount of nutrients in the litter should be sufficient to produce a response in the plants. This can be determined by appropriate calculations based on previous knowledge of crop demand, and nutrient content of litter. Pilot studies should determine litter decay rates, nutrient concentrations, and potential toxicity effects of the various litters. Treatments are:

1 Complete control (no amendments)
2 Low quality residue resource applied at 40 kg N/ha
3 High quality residue resource applied at 40 kg N/ha
4 Low and high quality resources each applied at 20 kg N/ha

High and low quality litter have narrow and wide C:N ratios, respectively. Low quality resources include such agricultural residues and waste products as cereal stover, sugar cane bagasse, sawdust and wood chips. These materials characteristically contain larger amounts of lignin and cellulose. Included among high quality resources are animal manures and other animal products, green manures, composts, agroforestry leaf litters and prunings and inorganic fertilisers.

1.3.4 Treatment rationale

The complete control (treatment 1) is compared to all treatments as a test of improved plant performance due to inputs. Treatment 1 also serves as the control treatment in all soil biological process measurements. Researchers must consider that this treatment may contain root and stubble residues.

The low quality resource treatment (treatment 2) is a measure of the potential immobilisation of the system and serves as a control when compared to the mixed residue quality treatment. Researchers should be aware that the quantities of this resource required to match the nutrient input levels of the high quality resource (treatment 3) are often excessive (> 10 t/ha) and may result in changes of other soil properties (e.g. soil water holding capacity, microclimate, etc.).

The high quality residue treatment (treatment 3) acts as a control of the nutrient use efficiency in the mixed treatment (treatment 4). Researchers must be aware that any root residues represent a low quality input common across all treatments and that the high quality residue input contains nutrients other than that which is limiting the growth of the complete control (treatment 1).

The low + high mixed litter quality treatment (treatment 4) is compared to treatments 2 and 3 as a test of the comparative nutrient use efficiency of the high quality input due to potential short-term immobilisation of plant nutrients resulting from the addition of the low quality input.

1.3.5 Experimental design

The experiment is installed as a randomised complete block design with 4 replicates (Figure 1.1). The dimensions of the individual plots are 10 m x 6 m. The distance between rows may be either 60 or 75 cm. This allows for either 10 or 8 crop rows per plot, the outside pair of which are not sampled. Because the residual effect of the treatments over time may be substantial, the study will be continued on the same plots for up to 3 years to assess these potential cumulative effects on plant growth and soil properties.

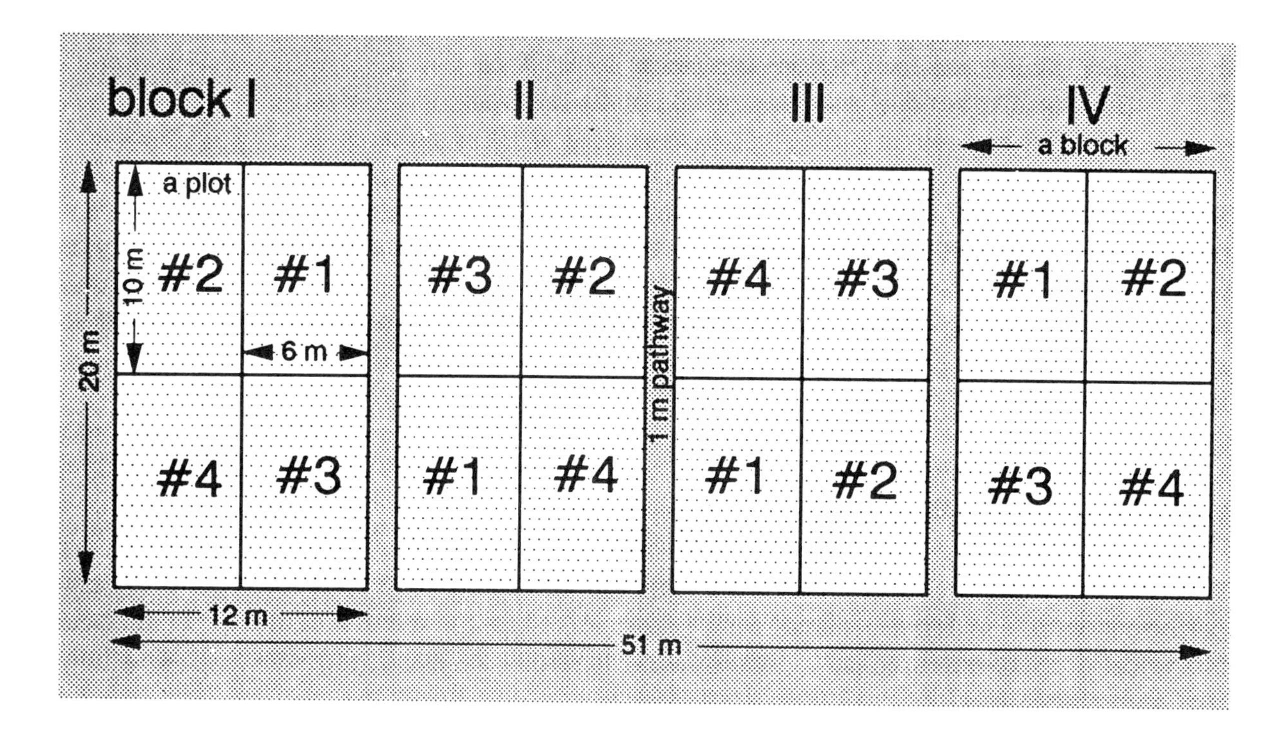

Figure 1.1 Experimental layout is a Randomised Complete Block design, with four replicates and four treatments. Treatments 1, 2, 3 and 4 are complete control, low quality input, high quality input and mixed inputs, respectively.

1.3.6 Experimental installation

The site should be mapped according to elevation and relevant soil characteristics. The entire field is tilled to a depth of 15 cm and the four blocks delineated based on the results of the initial site characterisation and terrain. Thus blocks may be irregularly shaped and non-contiguous if necessary to ensure uniformity of plots within a block. Then treatment plots (10 m x 6 m) are randomly placed within each block. The required additions to soils are evenly distributed across the soil surface and incorporated to 15 cm. Within the individual plots four areas are identified that will be sequentially sampled for above-ground biomass, root in-growth, litter decay, nutrient mineralisation and other soils properties. These sample areas are 1.5 m x 2.25 m. A final harvest area (4 m x 8.5 m) is also marked. A locally adopted cereal grain is planted at the recommended spacing and time of year.

As an example of this experimental approach, the rates of dry matter application of a low quality litter (e.g. sugar cane bagasse, 0.27% N) and a high quality litter (e.g. prunings from a nitrogen fixing tree, 2.7% N) to supply 40 kg of N/ha are presented in Table 1.1.

Table 1.1 The quantities of litter inputs required to provide 40 kg/ha.

Litter quality	Litter inputs (kg/ha)*	
	Sugar cane bagasse	Tree prunings
High	-	1,482
Low	14,815	-
High + Low	7,407	741
Complete control	-	-

* P, K, Ca, S, Mg, and micronutrients should be applied to ensure that only N limits crop growth.

1.3.7 Measurements

These should be repeated over time on the same plot.

1. Daily rainfall
2. Water and total N content of organic residue and soil
3. Soil physical characteristics
4. SOM fractions (total organic C, microbial biomass C, soil litter C)
5. Decomposition (litter bag technique)
6. N mineralisation (*in situ* and/or anaerobic)
7. Plant growth, e.g. total above-ground biomass for annual crops

8 Below-ground biomass
9 Crop yields
10 Nutrient content of above- and below-ground biomass
11 Population density of soil fauna, by functional groups, at the beginning and end of the study

1.3.8 Analysis and interpretation

The experimental design requires that this experiment be analysed through analysis of variance as a randomised complete block within sampling times and using regression techniques when different treatments are being compared over several time points. The comprehensive site characterisation and data collection also allows for the experimental results to be compared to CENTURY model simulations of several treatments.

1.3.9 Transition to on-farm experimentation

Representatives of non-governmental organisations, national agricultural laboratories and extension services should be invited to view the experiment towards its conclusion and discuss the experimental results. The most promising experimental treatments can then be installed into farmers fields in cooperation with interested individuals or organisations. It is advised that such follow-up experiments be researcher-installed and farmer managed, and that only crop yields and selected (few) soil parameters be measured at the conclusion of the experiment.

1.4 SITE CHARACTERISATION

The TSBF Programme is in essence a methodology designed to offer a novel approach to soil fertility research. A central feature of the TSBF approach however is the idea that in order to understand soil fertility dynamics in a given system, a minimum set of characterisation data are essential. A further key feature is to compare such data between research sites (this is greatly facilitated by sharing a common methodology) as it has been realised for some time that this would lead to a more rapid advancement of the understanding of soil biological processes, and improved management. Consequently, from the earliest stages of the Programme, the development of this Handbook has been seen as a priority. The idea of a minimum data set has evolved over the life of the TSBF programme, and continues to do so as techniques and understanding improve. The balance and scope of this "minimum package" have greatly benefited from input from a wide range of scientific disciplines, researchers working in sites with very diverse soils, climates and vegetation, and from the rigour of working in close collaboration with ecosystem modellers. The choice of parameters and methods chosen is guided by their suitability for laboratories in the tropics, robustness, cost and the amount of additional insight gained into processes operating within the ecosystem. The variables requiring determination for TSBF Research Sites are listed in Table 1.2. Variables were chosen so as to offer the minimum set of information required to make and interpret inter-site comparisons of soil biological processes at a gross level. Detailed studies may well require additional measurements not included in the table.

Table 1.2 Site characterisation measurements for TSBF research sites.

I Site Description

Present and past land use, topography and position on slope

II Site Variable Measurement

Variable	*Comment*	*Units*
CLIMATE		
Mean monthly precipitation		mm
Mean monthly maximum air temperature		°C
Mean monthly minimum air temperature		°C
SOILS		
USDA Soil Taxonomy	Family level	
Depth	To weathering parent material	m
then for 0 - 20 cm:		
pH	1:2.5 Water	
Organic carbon	Wet oxidation	%
Total N and P	H_2SO_4/H_2O_2/Se digestion	mg/kg
Exchangeable Al^{3+} (+ H^+) (if pH $<$ 6.0)	1 M KCl	me/100g
Exchangeable K^+, Ca^{2+} and Mg^{2+}	1 M NH_4OAc pH 7.0	me/100g
ECEC (if pH $<$ 6.0)	By summation of cations	me/100g
CEC (if pH $\geq$ 6.0)	1 M NH_4OAc pH 7.0/KCl	me/100g
Bicarbonate extractable P	pH 8.5	mg/kg
Resin extractable P	HCO_3^-/Cl^- resin	mg/kg
Organic P	Difference on ignition	mg/kg

Table 1.2 Continued

Micronutrients	If suspected to be limiting or toxic	mg/kg
Nitrogen mineralisation index	Anaerobic lab mineralisation	mg/g/d
Microbial biomass	Chloroform fumigation/extraction	mg/g
Soil litter	Flotation in water	mg/g
Mechanical analysis	Hydrometer: sand, silt, clay	%
Stone content (if > 5% by mass)		%
Field capacity	Gravimetric	%
Bulk density	Any suitable method	g/cm^3
VEGETATION		
Description and local name for system	e.g. Open Miombo woodland	
Woody plants by dominant species:		
Density		no./ha
Canopy mean maximum height		m
Basal area		m^2/ha
Herbaceous plants/crops:		
Peak standing crop	Harvesting @ 2 cm (state date)	t/ha
Species contribution to peak standing crop	Dry-weight-rank or sorted harvest	%
Roots		
Root profile	By counting	m/m^3
Root mass	Soil cores	kg/ha

Table 1.2 Continued

SOIL FAUNA		
Density	By functional group	no./m^2
Biomass (fresh weight)	By functional group	g/m^2
ORGANIC MATTER INPUTS		
Woody plant litter	Litter traps; reported monthly	kg/ha
Herbaceous litter/crop residue	Quadrats; reported monthly	kg/ha
Roots	[Specialist study]	kg/ha
ORGANIC MATTER INPUT QUALITY (For all inputs contributing > 10% of total input)		
Total N	H_2SO_4/H_2O_2/Se digestion	%
Polyphenolics (for leaves)	Folin-Denis	%
Lignin	*via* acid-detergent-fibre	%
DECOMPOSITION		
Decomposition standard	t_{50}	d
Tree leaf litter	Standard litter bags, t_{50}	d
Herbaceous litter	Standard litter bags, t_{50}	d
Litter turnover	By calculation	
SOCIOECONOMICS		
Socioeconomic context	Rapid Rural Assessment	

Table 1.2 Continued

if applicable to the system:		
HERBIVORY		
Animal type		
Animal mean live mass		kg
Stocking level	Per month	kg/ha
FIRE		
Date	Month of year	
Completeness	% Of standing crop burned	%
Intensity	From fire duration and fuel load	kJ/m/s

Chapter 2 SITE DESCRIPTION

2.1 CLIMATE

2.1.1 Climatic characterisation of site locality

To characterise the climate of a site, long-term data (>10 years and preferably, >20 years) are required, for at least monthly mean values of precipitation and temperature (maximum, minimum). Wherever possible information on soil temperatures, relative humidity (maximum, minimum), sunshine, wind (speed and direction) and evaporation should complement the "Minimum Dataset". These climatic data should be obtained from the nearest (or climatically most similar) weather station with an indication of the duration of record for each parameter and along with the particulars of the station such as name, latitude, longitude, altitude. Further information on the distance of the study site from that meteorological station, the relative altitudes, the aspects and topographical positions must be provided to enable an assessment of eventual climatic differences.

2.1.2 Meteorological observations

The site of a meteorological station should be representative of the natural conditions in the study area. It should:

1. be fairly level and free from obstructions;
2. have a short grass cover or natural cover common to the area (not be concrete, asphalt or rock);
3. not be closer to any obstruction (trees, buildings) than eight times that obstruction's height;
4. not be near areas with cold drainage, flooding or frequent sprinkling;
5. be easily accessible;
6. be fenced to exclude animals.

The daily observation times should be the same as used in the respective national meteorological service. If only one daily observation can be carried out it should be the morning observation. Minimum requirements for TSBF studies include:

Precipitation

Rainfall measurements should be made with standard gauges (WMO, 1983, Section 7) and the readings made to the nearest 0.2 mm. The orifice of the gauge must be horizontal and high enough to avoid splash from the ground, but low enough to prevent major wind influence. To assure comparability, standards as used by the national meteorological service should be followed. (Preference should be given to gauge types with at least 200 cm^2 receptive area.)

Air temperature

To obtain reliable air temperature data thermometers (dry bulb, maximum, minimum) or other probes used in connection with recording devices must be screened from direct radiation but yet be exposed to free air movement. Standard wooden or plastic screens, available for this purpose, are usually installed at a height of 1.2 - 2.0 m above the ground; the actual height chosen at a particular project site should coincide with the one used by the national meteorological service. The thermometers have to be calibrated and should have a graduation which allows a reading accuracy of 0.2°C.

Soil temperature

Standard soil thermometers should be positioned to measure the soil temperature at 0.10 m depth. The thermometers have to be calibrated and should have a graduation which allows a reading accuracy of 0.2°C.

Air humidity

Relative air humidity, defined as the ratio in percent of the actual water vapour pressure in the atmosphere to the one which can be held at a given temperature (saturation vapour pressure), can be directly measured with hygrometers or hygrographs. Absolute values of the vapour pressure (measured in hectopascals) are determined with psychrometers which consist of two thermometers; one is measuring the air temperature (dry bulb) while the other is covered by a wick kept moist with distilled, deionised water to depress the (wet) bulb temperature. The wet bulb wick must be kept clean and checked frequently for rot. The temperature difference between the dry and wet bulbs allows the calculation of the vapour pressure. As ventilation affects results considerably, models with forced ventilation (Assmann type) are strongly recommended. For recording purposes the Assmann psychrometer can be supplemented by a hygrograph.

Minimum air temperature near the soil

In high altitude areas which are liable to frost it is recommended to measure the minimum temperature near the ground. This is measured with a regular minimum thermometer exposed on a small shaded support 5 cm above the ground.

Sunshine

Solar radiation is the basic meteorological parameter for all heat and energy balance assessments. If no direct measurements of the solar radiation are made, the duration of bright sunshine allows for estimates of the energy available for physical and biological processes. The standard instrument to measure the duration of sunshine is the Campbell-Stokes recorder, where the focused radiation from the sun burns a trace in a chart. Sunshine duration is expressed in hours per day. As the sunshine patterns usually do not change over short distances, data from nearby meteorological stations may be used.

Wind

Wind data are essential for estimations of evapo(trans)piration. The basic instrument for measuring the wind speed is the cup anemometer; when equipped with a counting device it allows measurement of distances of wind run over time, and thus the mean wind speed over given time intervals can be derived. Data on wind speed should be complemented by information on the wind direction which can be observed with a wind vane.

In agricultural meteorology the standard height for wind measurements is 2 m above the ground but as topography and the surface cover influence the wind field to a large extent, observation sites have to be chosen very carefully.

Evaporation

To measure potential (maximum) evaporation directly, pans or open water tanks are exposed to the environmental conditions. The most commonly used evaporation pan is the Class A pan which has a diameter of 120.6 cm and a height of 25.4 cm. It is installed on a wooden grid approximately 13 cm above ground level, with a wire mesh protecting it from animals. The water level is controlled with a graduated hook gauge, set on a still well, or with a fixed point to which the pan is refilled with a calibrated container each time a measurement is made. The amount of water loss, analogous to rainfall, is expressed as the equivalent depth (in mm) of water evaporated.

Theoretical approaches, based on the surface energy budget, use measured data of sunshine duration, temperature, air humidity and wind run to estimate the maximum water loss from an open water surface, or from a surface covered with green grass well supplied with water (reference crop evaporation).

Recent research has shown that correlated data from shaded Piche evaporimeters provide good substitution for the aerodynamic term in formulas used for evaporation estimates. They would replace humidity and wind estimates.

Reference

WMO No. 8 (1983) *Guide to Meteorological Instruments and Methods of Observation.* Fifth Edition. World Meteorological Organisation.

2.2 SOIL CHARACTERISTICS

Data are needed at two levels:

1 Information necessary to ensure that the study site is representative of the region

This should be based on available regional soil maps coupled with reconnaissance survey if necessary.

2 Detailed information

Detailed information at a scale of 1:5000 or larger, is required for the actual site. The soils should be classified at the Family level of USDA Soil Taxonomy (Soil Survey Staff, 1987), and at least two (usually more depending upon the area) representative soil profile descriptions should be given (Breimer *et al.*, 1986) plus standard analyses (as listed in Table 1.2) of samples from 0-20 cm.

The soil parameters requiring measurement for TSBF can also be used to apply the Fertility Capability Soil Classification System (Sanchez *et al.*, 1982). This will help in the interpretation of soil constraints in terms of soil biological processes. Soil analytical methods are given in Chapters 6 and 7.

References

Breimer, R.F., van Kekem, A.J. and van Reuler, H. (1986) Guidelines for Soil Survey and Land Evaluation in Ecological Research. *MAB Technical Notes* **17**, UNESCO, Paris.

Sanchez, P.A., Couto, W. and Buol, S.W. (1982) The Fertility Capability Soil Classification System - Interpretation, Applicability and modification. *Geoderma* **27**, 283-309.

Soil Survey Staff (1987) *Keys to Soil Taxonomy* revised edition. Soil Conservation Service, Washington DC.

2.3 LAND MANAGEMENT

2.3.1 Land use

It is relatively rare to find sites on which the natural vegetation and soils have not been disturbed, to a greater or lesser degree, by land use. A description of the nature and intensity of such use is therefore an essential basis for the study of soil/vegetation dynamics and should be based on a classification plus a verbal description. The classification enables sites of similar nature to be related. The description enables account to be taken of the wide diversity of possible uses.

In the context of TSBF studies, it is particularly important to record the inputs and outputs to and from the ecosystem by the agency of man and domestic animals.

Area, site, present and past use

Land use is the use of land by man (including animals under control of man) for productive and/or service purposes. In the present context "use" should be interpreted widely, to include both land improvements (e.g. irrigation, drainage, terracing) and land degradation (e.g. forest clearance, salinisation). The spectrum of land use extends from the considerable disturbance of conditions through arable use at one end, to "undisturbed natural vegetation" as a limiting case at the other. Two distinctions are made: that between land use of the area and of the site; and that between present and past use.

The land use of the area refers to the surroundings within which the experimental site lies. Land within a circle of 5 km radius around the site may be given as a guideline. It will usually be necessary to describe the use in proportional terms, giving approximate percentages, e.g. "30% arable (mainly cereals and legumes), 20% pasture and 50% natural woodland, used for grazing and collection of fuel wood and domestic timber". The land use of the area does not directly concern the dynamics of the site, but provides a setting, enabling it to be recognised as broadly similar to that in other countries. For example, "former shifting cultivation, now changed by population pressure into semi-permanent arable use" indicates a situation of widespread applicability.

The site refers to use on the actual site of the experimental work. Land use will normally be the same over the whole site. Present land use is the use during the immediately preceding year. It will frequently include differences in use as between wet and dry seasons, e.g. "maize for 4-5 months commencing in the wet season, natural fallow with limited cattle grazing for the rest of the year". Land clearance, cultivation (hand hoeing, animal or tractor-powered ploughing, etc.), sowing/planting and harvesting methods should be recorded with date. The full description of farming systems is a substantial task and the subject of a wide range of publications (e.g. Simmonds, 1985).

Past land use refers to the trends of use in recent years. For the area as a whole, this provides a further element of contextual setting. For the site, the intention is to cover whatever period may have appreciably affected the present condition of the soils and vegetation. A 10-20 year span should be included as a minimum, with brief reference to earlier land use history where appropriate, e.g. "Forest clearance in this area is thought to have taken place about 50-60 years ago; for the past 10 years, the site has been under semi-natural pasture, intensively grazed by cattle, sheep and goats".

Classification (after Young, 1985):

1 Annual crops
Arable use for annual, or quasi-annual, crops. Include cassava, hill rice, vegetables if on a field scale; exclude swamp rice, gardens.

2 Swamp rice
If irrigated, list also as irrigation.

3 Tree and shrub crops
Perennial crops, excluding field perennials.

4 Field perennials
Sugar cane, sisal, pineapple, bananas.

5 Gardens
Intensive production, of vegetables and/or fruit, on small plots.

6 Irrigation
Include rice if water brought to fields, but not if retention of rainfall only; include irrigated pasture, forest.

7 Natural pasture
Livestock production from natural pasture, including nomadic grazing, ranching.

8 Improved pasture
Livestock production from substantially improved or sown pasture.

9 Forestry: natural forest
For timber and/or other products.

10 Forest plantations
For timber and/or other products.

11 Agroforestry
Trees interacting with crops and/or pastures/livestock.

12 Wildlife conservation
Flora and fauna, intentional use for this purpose.

13 Water catchments
Intentional use for this purpose.

14 Engineering use
Any form of construction.

15 Unused
Natural vegetation, with no significant use.

Format

The following format for land-use classification is suggested:

Area Present use:
Classes (with approximate percentages):
Description:
Past use:
Description:

Site Present use:
Class:
Description:

Management: (e.g. methods of vegetation clearance, sowing/planting, weeding, pruning, harvesting)
Inputs: (types, quantities, composition, timing and manner of application)
Organic material:
Inorganic material:
Water (irrigation):
Outputs: (harvest etc.; types, quantities, composition)
Past use:
Description:
Inputs and outputs: (general account, approximate quantities where possible)

References

Simmonds, N.W. (1985) *Farming systems research: A review.* World Bank Technical Paper No. 43. World Bank, Washington DC.

Young, A. (1985) *An environmental database for agroforestry.* ICRAF Working Paper 5 (revised edition), ICRAF, Nairobi, Kenya.

2.3.2 Fire

The required data are the date, completeness and intensity of the burn. Completeness is a visual estimate, or the difference between a before and after fire sample, of the percentage of the litter, herbaceous and woody standing crops consumed by the fire, and thus estimates of carbon loss. Intensity is a measure of the rate of energy release by the fire related to nitrogen and phosphorus losses. It can be calculated from the fuel load (F, kg/m^2), the energy content of the fuel (E, kJ/g: for most biomass this can be assumed to be 20 kJ/g), and the rate of spread (V):

$$V\ (m/s) = A\ /\ (w \times t)$$

$$Intensity\ (kJ/m/s) = F \times E \times V$$

where

A = area of plot (m^2)
w = width of flame front (m)
t = time to burn plot (s).

Alternatively, intensity can be estimated from the following rough guide (Trollope, 1984):

Category	Intensity (kJ/m/s)	Description
Warm	> 500	Flames < 1 m, easily approachable
Hot	500-1000	Flames 1-1.5 m, approachable
Very hot	1000-2000	Flames 1.5-3 m, difficult to approach
Extremely hot	>2000	Flames > 3 m, uncontrollable

In keeping with the peak-trough philosophy of vegetation monitoring, herbaceous biomass (fuel load) should be assessed before burning.

Reference

Trollope, W.S.W. (1984) Fire Behaviour. In: Booyessen, P. de v. and Tainton, N.M. (eds.) *Ecological Effects of Fire in South African Ecosystems. Ecological Studies* **48** Springer-Verlag, Berlin.

2.3.3 Socioeconomic components

The TSBF programme recognises the need to focus local research programmes on local land use practices. The overall objective of TSBF is to determine management options for improving tropical soil fertility through the manipulation of biological processes. Within the range of more specific TSBF objectives, there is a need to choose those that have local applicability.

There are different approaches and levels of investigation required in a program of research on soil fertility. Basic research, or strategic research, on soil biological fertility provides the information needed for target research and extension programs assisting farmers in coping with fertility constraints such as low productivity due to soil erosion. The application of the principles learned from strategic research to the local soil fertility problems as defined by the target population requires an understanding of the socioeconomic context of the farming systems in which they operate.

There are many techniques used to gather socioeconomic data. One of these is participant observation. Participant observation allows the researcher to gather an in-depth knowledge of the indigenous farmers' practices and beliefs. It can be quite time consuming and is better suited to smaller populations. A second technique is the use of questionnaires which are standardized, structured and suited to quantifying socioeconomic data of generally large populations (for example, see Moran in TSBF Handbook of Methods, first edition). A standardised questionnaire cannot respond to unanticipated site-specific problems (Beebe, 1985) and the analysis can be quite time consuming. A third technique is in-depth interviews which are used to gather data using clearly specified criteria generally aimed at a better understanding of specific problems or processes. This technique uses a flexible data collection/processing procedure, generally focuses on smaller populations (L. Llambi, Caracas, Venezuela, 1988, personal communication), and can be adjusted to time constraints and site-specific problems. TSBF researchers used a structured questionnaire, proposed in the 1986 TSBF Yurimaguas Workshop by Moran (1986). They concluded that structured questionnaires were not expedient in performing multidisciplinary land use surveys. It was proposed that in-depth interview techniques such as Rapid Rural Appraisals which pay attention to context be used as effective procedures for recognizing constraints and opportunities for land users.

Rapid Rural Appraisals were designed to use observation and in-depth interviews to gather data on agricultural systems in developing countries. Rapid Rural Appraisal (RRA) is an interdisciplinary, iterative approach used to gain insight into land use practices. Small teams of researchers representing different disciplines such as social science and soil science work together using techniques such as in-depth interviews, based on key indicators, and direct observation to gain insight into management practices. It is a highly iterative approach which allows the reformulation of hypotheses based on new information (Grandstaff *et al.*, 1987). The real value of techniques such as RRA are the use of indigenous technical knowledge (ITK) such as the farmers' knowledge of plants, soils, and the local ecosystem to reformulate these hypotheses. The relevance and validity of these rapid techniques are evaluated in the short term through repeated iterations of the RRA process and in the long term by the effectiveness of actions based on knowledge and perceptions generated by the RRA.

Organisations such as ICRAF (Raintree, 1987) and IBSRAM (1988) are successfully using techniques such as RRAs to assess soil management techniques and other agroecosystem concerns in developing countries.

An example of an RRA is given in Appendix A. The decision to choose between participant observation, standardised questionnaires, and in-depth interviews is dependent on the timing of the project (Murphree, 1989) and the need for flexibility vs. standardised quantifiable data. If time is not an issue, participant observation or standardised questionnaires are possible. In-depth interviews can also be used as a starting point with the standardised questionnaire being used as the follow up. When time is an issue, in-depth interviews combined with observation and repeated iterations can be modified to suit the particular needs of a project such as TSBF. Ultimately the intended use of the data and the time allotted for data collection should guide the choice of data collection techniques.

References

Beebe, J. (1985) *Rapid Rural Appraisal: The Critical First Step in Farming Systems Approach to Research.* Farming Systems Support Project, Networking Paper No. **5**, USAID, Washington, DC.

Grandstaff, S.W., Grandstaff, T.B. and Lovelace, G.W. (1987) Draft Summary Report. In: *Proceedings of the 1985 International Conference on Rapid Rural Appraisal.* Khon Kaen, Thailand: Rural Systems Research and Farming Systems Research Projects.

IBSRAM (1988) Methodological guidelines for IBSRAM Soil Management Networks. Second draft, April. IBSRAM, Bangkok, Thailand.

Moran, E.F. (1986) TSBF Inter-regional research planning workshop. In: Swift, M.J. (ed.) *Biology International* Special Issue **13**, IUBS, Paris, 53-64.

Murphree, M. (1989) Socio-economic research components in tropical soil biology and fertility studies. *Biology International* Special Issue **20**, 31-40.

Raintree, J.B. (1987) *D and D User's Manual. An introduction to Agroforestry diagnosis and design.* ICRAF, Nairobi, Kenya.

Chapter 3 FIELD PROCEDURES

Note: For *in situ* N-mineralisation, see Section 6.7.2.
For *in situ* soil physical determinations, see Chapter 7.

3.1 ABOVE-GROUND VEGETATION

The methods described below are recommended for use by researchers engaged in TSBF studies for site description. Intensive sampling should be confined to the actual study site, but the surrounding vegetation should be described in general terms (extent, dominant species, overall structure, ecological affinities etc., and local name, vegetation type) making particular note of past management history (e.g. age of stand, fire history, grazing etc.). A general text covering vegetation description is Mueller-Dombois and Ellenberg (1974).

Reference

Mueller-Dombois, D. and Ellenberg, H. (1974) *Aims and Methods of Vegetation Ecology.* Wiley, New York.

3.1.1 Woody vegetation

The desired information is the above-ground woody biomass, but since it is seldom practical to measure this directly, an indirect approach using the stem basal area (m^2) and height (m) of a sample of woody plants is used. In order to extrapolate the sample to a per hectare basis, the total basal area per species per hectare (m^2/ha) and the plant density (plants/ha) are also required.

Species

Species should be identified by their full scientific name (*Genus species* Author). Unidentified species should be given a code or local name, and a reference to a specimen in a herbarium.

In natural communities with a large number of species, it is not necessary to record the height and basal area of minor species individually. All the species contributing less than 10% of the total biomass can be combined into one category.

In extremely diverse communities with no clear biomass dominance, where no species contributes more than 10% to the total biomass, it may be necessary to define the "species" on a functional basis; for instance emergents, canopy trees, sub-canopy trees and understorey shrubs, recording which species were placed in which category.

Stem density: (i) Plot methods

If the trees are sparse, then all the trees within the study site can be individually counted and measured. Usually this is not the case, and one or more subsamples including about 100 trees are taken.

A square quadrat of known area can be laid out with the aid of a tape measure, a compass and string, but often a circular quadrat (all the trees within a given distance of a centre peg) is easier to work with.

A long thin quadrat (belt transect) is often an efficient method in dense vegetation. Lay out a tape measure (50 m) in a straight line, and record all the trees within a given distance (1-5 m, measured with a hand-held rod) on either side of the tape. Both the length and width of the transect can be varied to give a total sample of about 100 plants; different widths can be used for common shrubs and sparse trees within the same sample. Individuals (or multistemmed clumps) more than half within the plot boundary should be counted in. To permit biomass estimates based on individual trees at a later stage, data should be recorded as in the following example:

Plots are 5 m x 50 m

Species	Height (m)	Diameter (cm)	Stem area (cm^2)
Burkea africana	3.2	15	176
Ochna pulchra	1.5	5	19
Sclerocarya birrea	6	30	707

Stem density: (i) Plotless methods

Laying out plots is time consuming and can lead to error due to subjective placement of the plots. A variety of plotless methods are available, such as the Point-Centre-Quarter method (PCQ; Cottam and Curtis, 1956). These methods use the distance between random points and the nearest plant to calculate an average inter-plant distance (d_x). The number of plants per hectare (N) is then given by

$$N = 10,000 / d_x^2$$

where d_x is in metres.

About 100 distance measurements are required, and at the same time the species, stem diameters and heights of this random sample of plants are recorded.

The PCQ method is not applicable to vegetation which is regularly spaced (such as plantations) or highly clumped; various modifications, such as the two-distance methods of Diggle (1975) or Cox (1975) must be used.

Height

Height is the vertical distance between the ground and the highest living part. For trees taller than 5 m an estimate to the nearest m is sufficient; smaller trees should be recorded to the nearest 0.1 m. A measuring rod is the easiest method in low vegetation. In tall vegetation, where the top and the bottom of the tree can be simultaneously observed, a clinometer can be used; otherwise any practical method can be applied.

The mean maximum height for each vegetation layer (stratum) should be estimated from the data. For example, for a primary tropical lowland forest the emergents may reach 60 m, the canopy trees 40 m, the subcanopy trees 20 m and the understorey shrubs 6 m.

Stem basal area

The stem basal area (A) is the cross-sectional area of the trunk (m^2) measured at the lowest point not influenced by basal swellings or buttresses. In forests it is traditionally measured at 1.3 m above ground level ("diameter at breast height", dbh), but in tropical forests the trees may still be buttressed at this height, and in savannas, branching frequently occurs below this height. The area can be calculated either from a diameter (d) measurement using callipers, or from a circumference (c) measure using a tape.

$A = 0.7854\ d^2$ or
$A = 0.0796\ c^2$

The result will be in m^2 if the diameter and circumference are measured in m; if measured in cm, divide the answer by 10,000 to convert cm^2 to m^2. If the stems are very markedly non-cylindrical (for instance if they are elliptical or convoluted) an empirical correction factor should be calculated.

Where a single plant has more than one stem, the area of each stem should be measured. For shrubs with very many small stems an estimate of the total combined area is adequate.

The total stem basal area for the plot is calculated by multiplying the stem basal area by stem density for each species (according to the criteria for species as defined above).

Tree biomass estimation: natural vegetation

Tree biomass is most precisely estimated by cutting and weighing all portions of a tree. As this is very time consuming, tree biomass is more commonly estimated by statistical means, e.g. by sub-sampling, or on estimates based on tree attributes. When sub-sampling is not possible, biomass is most commonly estimated from a regression equation.

Since mass is the product of volume and density, tree biomass regressions often use diameter at breast height (D, cm), total height (H, m) and some measure of specific wood density (S, t/m^3). A locally developed equation is generally most reliable, and should be used whenever possible. When no local equation is available, one of the regression equations given below can be used. These have been derived from several data sets and rigorously tested for reliability (Brown *et al.*, 1989; Brown and Iverson, 1992; Brown, 1996). Equations using D, H and S are the most reliable, followed by those using D and H, or just D. When H is not available one may apply double sampling methods, and estimate H as a function of D, then biomass as a function of D and estimated H. This is only better than approaches based on D alone when the H-D relationship accurately reflects the population of interest.

The regression relationships vary slightly with climate, so there are three general sets of equations reflecting three rainfall regimes: dry <1500 mm, moist 1500-4000 mm and wet > 4000 mm. Not all equations are available for all climate types, due to lack of reliable data. The "dry" rainfall regime has been divided into two because of the strong influence rainfall has on tree biomass in this zone. Equations may be applied to individual trees, mean trees or stand tables (numbers of stems per size classes) to estimate total above ground biomass for a given tree, for a mean tree or per unit area respectively (cf. Wharton and Cunia, 1985, for further details of methods). The following equations estimate biomass and total height for trees in tropical forests. Values for the range of diameters included in datsets, sample size, n, and goodness of fit (adjusted R^2) are included. It should be stressed that although these equations have been derived from large (and sometimes very large) sample sizes, they should be viewed with some caution until validated for a given system.

Biomass regression equations for estimating biomass of tropical trees. B, biomass per tree (kg); D, dbh (cm); BA, basal area (cm^2); and H, height (m).

Forest type	Equation	Range in dbh (cm)	Sample size	Adjusted R^2
Dry	$B = \exp\{-1.996 + 2.32 * \ln(D)\}$ [a]	5-40	28	0.89
	$B = 10\{-0.535 + 0.966 \log_{10}(BA)\}$ [b]	3-30	191	0.94
Moist	$B = \exp\{-2.134 + 2.530 * \ln(D)\}$	5-148	170	0.97
	$B = \exp\{-3.1141 + 0.9719 \ln(D^2H)\}$		168	0.97
	$B = \exp\{-2.4090 + 0.9522 \ln(D^2HS)\}$		94	0.99
	$H = \exp\{1.0710 + 0.5677 \ln D\}$		3824	0.61
Wet	$B = 21.297 - 6.953(D) + 0.740(D^2)$	4-112	169	0.92

Note: exp{...} means "e to the power of {...}"
10{...} means "10 to the power of {...}"

[a] Revised from Brown *et al.* (1989) and recommended when precipitaion = 900-1500 mm/yr.
[b] From Martinez-Yrizar *et al.* (1992) and recommended when precipitation <900 mm/yr.

Tree biomass estimation: even-aged plantations

Woody biomass estimation in even-aged plantations is simpler than in natural forest, and requires destructive sampling of many fewer trees. The same principle, of estimating biomass from predictor variables such as stem cross-sectional area and height, applies, but the model relating these variables to biomass cannot be assumed to be the same for different species and trees of different sizes.

For many species, for a given stand structure, the following linear function gives an accurate prediction of biomass (Stewart *et al.*, 1992):

$$w = a_{ij} + b_{ij}\Sigma d^2$$

where

w = biomass
d = stem diameter (Σd^2 being the sum of the squared diameters of all the stems of a multi-stemmed tree)
a and b are coefficients for species i on site j.

These coefficients have to be determined separately for each species and each site.

Procedure

1 Fell a minimum of 12 trees.

2 Partition the trees into components (wood of different sizes, leaves, and pods/fruit).

3 Weighed the components immediately in the field.

4 Oven dry subsamples (> 100 g) of each to determine water content.

5 Calculate total dry weight (biomass):

The biomass and diameter(s) of the sample trees are used to determine a and b by regression analysis. If R^2 is low (< 0.8) other models should also be tested. The biomass of the other trees in the plantation is then estimated from their diameters. For a small stand this may be done for every tree individually; for larger populations the total biomass may be estimated using the mean diameter and the total number of stems in the stand. Twelve is the absolute minimum number of trees per species that must be destructively sampled, but if a larger sample is used it will give a more reliable model with closer confidence limits. Whatever the sample size chosen, it is essential that the sampled trees cover the whole size range of the population. The usual procedure is to use a stratified sample in which the trees are divided into size classes on the basis of diameter (or cross-sectional area in the case of multi-stemmed trees), and sample trees are selected randomly within each size class.

For studies on a single site, the inclusion of tree height h in the regression function is often found to contribute little to the goodness of fit of the model. However, a linear model of the form

$$w = a_i + b_i h\Sigma d^2$$

is more likely to be applicable across sites than one using diameters only, because height, unlike diameter, can be assumed to be independent of stocking level and silvicultural treatment (Philip, 1983).

References

Brown, S., Gillespie, A.J.R. and Lugo, A.E. (1989) Biomass estimation methods for tropical forests with applications to forest inventory data. *Forest Science* **35,** 881-902.

Brown, S. and Iverson, L.R. (1992) Biomass estimates for tropical forests. *World Resources Review* **4,** 366-384.

Brown, S. (1996) A primer for estimating biomass and biomass change of tropical forests. FAO Forestry Report, (in press).

Cottam, G. and Curtis, T. (1956) The use of distance measures in phytosociological sampling. *Ecology* **37**, 451-460.

Cox, T.F. (1975) The robust estimation of the density of a forest using a new conditioned distance method. *Biometrika* **63,** 493-499.

Diggle, P.J. (1975) Robust density estimation using distance methods. *Biometrika* **62,** 39-47.

Martinez-Yrizar, Sarukhan, A.J., Perez-Jimenez, A., Rincon, E., Maass, J.M., Soli-Magallanes, A., Cervantes, L. (1992) Above-ground phytomass of a tropical deciduous forest on the coast of Jalisco, Mexico. *Journal of Tropical Ecology* **8,** 87-96.

Philip, M.S. (1983) *Measuring trees and forests.* Tanzania/Dar-es-Salaam. Division of Forestry, University of Dar-es-Salaam.

Stewart, J.L., Dunsdon, A.J., Hellin, J.J. and Hughes, C.E. (1992) *Wood Biomass Estimation in Central American Dry Zone Species.* Tropical Forestry Paper **26**, Oxford Forestry Institute, Oxford. 83pp.

Wharton, H.E. and Cunia, T. (1985) *Estimating tree biomass regressions and their errors.* USDA Forest Service: Northeastern Forest Experiment Station, Broomall, PA. Report NE-GTR-117.

3.1.2 Herbaceous plants and short duration crops

The required measurements are total biomass (t/ha = 0.001 x kg/ha = 0.01 x g/m^2) and the contribution (%) of each species to that total. Total biomass is measured by harvesting, drying and weighing a number of small subsamples, or quadrats. Quadrat size depends on plant spacing: 0.5 m x 0.5 m is a convenient size for most grasslands; 1 m x 1 m may be appropriate for crops such as maize. Sample number (n) should be sufficient to reduce the standard error to about 10% of the mean. 20 - 30 samples per treatment, distributed amongst replicates, is usually sufficient. Sample location should best be random, but systematic with a random start is acceptable.

Procedure

1 Cut all herbaceous vegetation within the quadrat at 2 cm above the ground (to avoid soil contamination), and sort into live (green) biomass and standing dead if possible.

2 Collect the litter from the ground for an estimate of litter standing crop.

3 Dry all samples as soon as possible to prevent decomposition.

Species composition in mixed communities is estimated by the dry-weight-ranking technique. The technique is based on a multiple regression for the dry weight of a mixed sample of herbage on the weights of the three heaviest species in the mixture. Experience indicates that it is easier to assess visually the rank order of the species in a quadrat than to estimate accurately their biomass. Tests in a large number of different communities have shown that the regression coefficients are fairly consistent between communities, and therefore do not need to be recalculated each time. The technique does not work well in communities completely dominated by one species, and tends to ignore rare species. The modified form given here, where the total biomass within the quadrat is also given a visual score, gives better results in communities where the total biomass is patchily distributed. Quadrat size should be small enough that species ranking is simple, but large enough that most quadrats have at least three species in them. Quadrats 0.5 m x 0.5 m square are usually adequate in grasslands. About 50 quadrats should be assessed per treatment.

Procedure

1. Walk around the plot to obtain a clear visual impression of what the minimum (1) and maximum (5) quadrat biomass looks like.

2. Locate the quadrats randomly or systematically after a random start.

3. In each quadrat (i) give the total biomass a score (w_i) between 1 and 5 according to whether it is near the minimum or maximum for the plot.

4. In each quadrat, give the species (j) which contributes most to the total quadrat biomass a rank score (r_{ij}) of 1, the second heaviest species a rank of 2, and the third heaviest species a score of 3. If a single species contributes more than about 70% of the biomass, give it ranks 1 and 3, or 1 and 2 (or even 1, 2 and 3 if it is the only species in the quadrat). Similarly the second species could get a 2 and 3 if necessary.

5. When all the quadrats have been scored and ranked, calculate the score for each species:

Calculation

$$\text{Species}_j \text{ score} = 70.2_r\Sigma w_i + 21.1_r\Sigma w_i + 8.7_r\Sigma w_i$$

where
Σw_i is the sum of the quadrat scores for the quadrats where species j obtained rank r.

Add up the species scores to give a total score.

Determine the % contribution by species j to the total biomass:

Species j contribution to biomass (%) = (species score / total score) x 100.

Further reading

Gillen, R.L. and Smith, E.L. (1986) Evaluation of the dry-weight-rank method for determining species composition in a tallgrass prairie. *Journal of Range Management* **39**, 283-285.

Sandland, R.L., Alexander, J.C. and Haydock, K.P. (1982) A statistical assessment of the dry-weight-rank method of pasture sampling. *Grass and Forage Science* **37**, 263-272.

3.2 ABOVE-GROUND INPUTS

3.2.1 Tree and shrub litter

In a comprehensive review of tropical litter fall data, Proctor (1983) observed that the results of many published studies were not comparable. This resulted from inadequate siting and replication of traps in relation to site heterogeneity, sampling for periods of less than a year and lack of standardisation of small litter fractions (leaves, twigs, reproductive structures and "trash").

Litter trap construction

Litter traps are bags or boxes supported just clear of the ground with an aperture of 0.25 - 1 m^2. A circular construction is best as it minimises edge effects. Woven plastic bags are light weight for use in remote sites and can be tensioned into shape using lines attached to D-rings sewn around the mouth of the bag. The traps must allow free drainage of rain water but have a mesh size of approximately 1 mm or less to retain fine litter fractions. Trays on the ground surface can be used to measure litter-fall from dwarf shrubs etc., but animal activity, drainage and wind can present problems.

Similar considerations apply to collections of litter from quadrats on the ground, which may be necessary for estimating falls of palm fronds and larger woody litter. Trash fractions, however, which often have low mass but high nutrient content, will be lost by this method.

Procedure

1 Randomly locate litter traps (for material other than branches) within moderately homogeneous plots, or in a stratified random pattern (with 10 traps per subplot) in sites where it is necessary to include major variation in topography, soils and vegetation structure. **Note:** To achieve a 5% standard error about the mean, Newbould (1967) recommends the use of at least 20 traps/plot. In very heterogeneous sites, however, higher numbers of traps may be required.

2 Collect litter every 2 weeks and air dry it. More frequent collections may be necessary for litters which decompose rapidly, e.g. some tree legumes, while less frequent collections may be made under dry conditions (though the possibility of the litter becoming contaminated with dust and/or animal faeces should be recognised).

3 Sort the dried material into:

a) leaves (including petioles and foliar rachises);

b) small woody litter (twigs < 2 cm in diameter and bark);

c) reproductive structures (flowers and fruits could be differentiated);

d) trash (sieve fraction < 5 mm).

(For palm fronds, the leaflets, the rachis below 2 cm, and the remaining parts of the rachis should be weighed and recorded separately.)

4 Oven-dry subsamples of litter to obtain correction factors for moisture content (see Section 6.1)

5 Express all fractions defined above on an oven dry basis in g/m^2/year or t/ha/year with 95% confidence limits.

6 Estimate branch fall from large (e.g. 100 m^2) ground quadrats. Break twigs at the 2 cm diameter point, weigh the > 2 cm diameter material, subsample for oven-dry mass and other determinations as required.

References

Newbould, P. J. (1967) *Methods of Estimating the Primary Production of Forests.* Blackwell Scientific Publications, Oxford.

Proctor, J. (1983) Tropical Forest Litterfall 1. Problems of data comparison. In: Sutton, S.L., Whitmore, T.C. and Chardwick, A.C. (eds.), *Tropical Rain Forest: Ecology and Management.* Blackwell Scientific Publications, Oxford, pp.267-273

3.2.2 Herbaceous litter and above-ground crop residues

The minimum level of sampling is at maximum and minimum biomass associated with major seasonal changes or perturbations; i.e. sampling four times a year under climatic regimes with a strongly bimodal pattern of rainfall. This will underestimate litter inputs as a consequence of material turning over between sampling dates and sampling at regular intervals every few weeks is recommended.

Procedure

1 Determine herbaceous litter (including grasses and forest ground flora) by harvesting quadrats in conjunction with biomass estimates (Section 3.1.2).

2 Separate litter, where possible, by plant species for the most frequent 80% of species and bulked for the remaining 20%. (This may be impractical in very species-rich communities.)

3 Determine crop residues after harvest and at the time of ploughing or other manipulation.

4 Oven-dry litter subsamples to obtain correction factors for moisture content (see Section 6.1). Litter heavily contaminated by soil may need to be corrected for 'ash' content as well (see Appendix D 2.6).

5 Express all fractions defined above on an oven dry basis in g/m^2/year or t/ha/year with 95% confidence limits.

3.2.3 Other inputs

Mulches, manures and fertilisers

Information on the type of input applied, the amount applied, and the date of application is necessary. For example, information required for mulches would include the type of mulch (e.g. rice straw), the amount applied on a dry weight basis (t/ha), and the application date(s). A representative subsamples can be taken for nutrient and other analyses. Details on methods of incorporation (e.g. surface application, ploughed to a certain depth, banded) of each of the inputs should also be noted. Similar information is necessary for manure application.

For fertilisers, specify the material applied (e.g. urea, single superphosphate), quantity applied and application date(s).

Herbivory

Record the dates of commencement and termination of feeding, the type of herbivore (species, browser, grazer, etc.), the number of animals in the plot and estimates of their individual masses. These data will permit the stocking level (kg/ha) to be calculated, given the plot area, and the rates of plant consumption, defecation and urination (kg/ha/day) to be estimated.

3.3 ROOTS

Note: This section gives procedures for the study of the overall root pattern. A review of methodologies for estimating root biomass, production and estimating total root carbon input to the soil may be found in Appendix D ("Roots: length, mass, biomass, productivity and mortality").

Root research requires destructive sampling of the soil, often causing considerable disturbance to the plots. Space should be allowed for this when designing field experiments. Specific information of root systems (both vertical and horizontal) is important when trying to understand plant/input interactions and the correlative effects on soil fauna and microorganisms. Information on the lateral spread of root systems is essential in deciding on "guard" areas or borders in field experiments. Errors in interpreting results are easily made when no root information is available, as lateral spread of over 5 m (cassava) or up to 20 m (certain trees) is often more than expected. Information on rooting depth and distribution is also essential when placement effects are to be considered regarding plant nutrient uptake.

Apart from the two classical descriptions of root methods by Schuurman and Goedewaagen (1971) and Böhm (1979), a number of recent reviews is available on methods to quantify root development and functioning in the field: Caldwell and Virginia (1991), Mackie-Dawson and Atkinson (1991), Taylor *et al.* (1991), van Noordwijk (1987), van Noordwijk *et al.* (1992).

Three methods outlined below are based on the study of soil profiles from a soil pit. In soils without stones or woody roots a monolith sampler as described by Floris and van Noordwijk (1984) might have the advantage of less site disturbance, but generally soil pits are needed to have access to the root environment.

3.3.1 Root preparation on profile walls

A soil pit is dug close to the plant selected for study and by carefully removing soil close to the stem some main roots are identified; their course is followed by gradually removing the surrounding soil, using a pin. When all (major) roots within the first, say, 10 cm from the original profile wall have been exposed, a drawing is made e.g. on a 1:5 scale on graph paper (Figure 3.1), using pins to mark grid points in the soil. Unfortunately most fine roots will break off during this procedure, and so only a qualitative picture is obtained. Still, the method allows width and depth of the root system and branching pattern to be recorded. The response of the root system to heterogeneities of the soil, to transitions between soil layers, to cracks (clay soils) and channels in the soil (made by soil fauna and/or previous roots) deserves special attention.

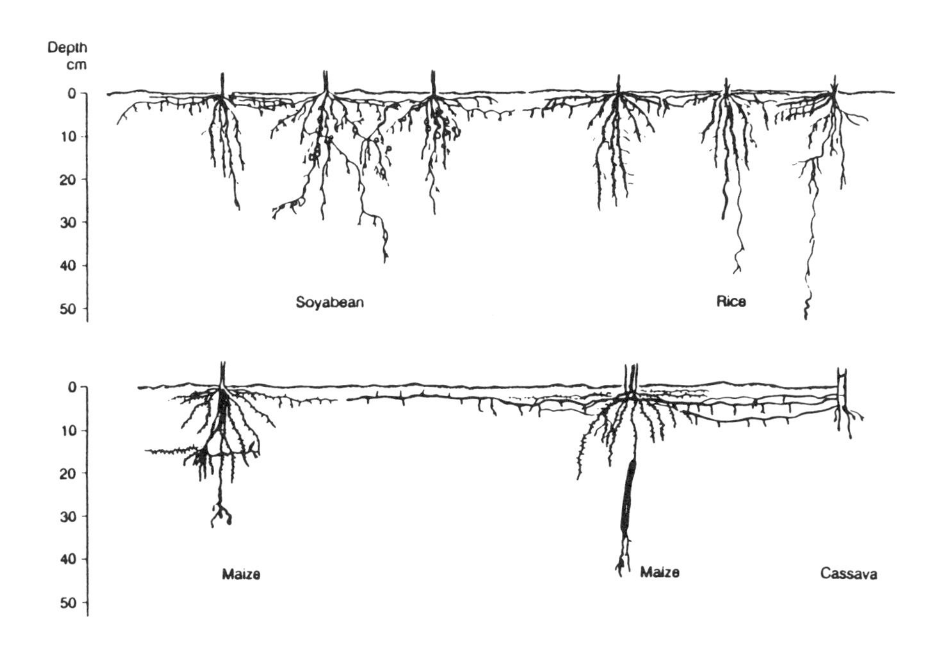

Figure 3.1 Example of root observations on profile wall, showing the response of maize roots to an acid subsoil (below 15 cm) and to the presence of old tree root channels (van Noordwijk *et al.*, 1991a).

In mixed cropping situations roots of the various components can usually be recognized after some training. By spending 1 to 2 days per major species the overall patterns of shallow and deep rooted species and lateral spread can be estimated: both are important when designing fertilizer or crop residue experiments or when designing mixed cropping systems. Results of this method are easily understood and can be used in discussions with e.g. farmers.

For shallow rooted crops lateral roots can also be observed from the stem base by digging a small trench, following the root. The presence of "sinker" roots, vertically oriented branch roots from a horizontal branch root, is of special interest here. By excavating a small area around the stem a classification of roots by diameter and orientation can be made (van Noordwijk *et al.*, 1991b). This way replicated observations can be made and a "typical" specimen can be selected for further observation.

3.3.2 Root mapping on profile walls

Based on results of root pattern observations (described above) an estimate of total root density in the profile can be obtained by root mapping on polythene (PVC) sheets. Profile walls are cleaned and straightened with a sharp knife, after having removed stones and thick roots which would hinder the process. Depending on soil texture, various steps can be taken to facilitate root observations:

- on sandy soils a water spray (knapsack sprayer) can be used to remove about 2 mm of soil to expose roots,

- on clay soils gently brushing the soil may help. Good results have also been obtained covering the profile wall overnight with a cotton cloth soaked in a 5% sodium hexametaphosphate solution to disperse the clay prior to spraying. In dry soils compressed air can also be used to blow away crumbs of soil; roots are remarkably resistant to such treatment.

After preparation the profile is covered with a transparent PVC sheet and carefully searched for roots; a predrawn 10 cm x 10 cm grid on the PVC helps to work systematically. All roots are marked with dots on the sheet; differently coloured pens can be used for different size classes or plant species. Roots are only recorded where they intercept the plane of observations; branch roots outside that plane, exposed by preparation of the profile wall, are neglected. Major features in soil structure and horizon boundaries should also be noted (Figure 3.2 a).

(As condensation build-up behind the PVC sheet may present a problem in some instances, a wire (or fine string) grid may be used: the grid is made on a light-wooden frame (1 m x 0.5 m), with the wires stretched across at 10 cm spacings. (A suitable wire mesh could be used.) The coordinates are noted rather than marking the PVC sheet. A comparative advantage of this method is that it allows a closer examination of the root condition and soil features; a detailed spatial patterning is, however, harder to record.)

Root intensity (number of interceptions, N, per unit map area) can now be estimated for each soil horizon, as a function of distance to the plant. A more detailed analysis of root pattern (regular, random or clustered) within each horizon is possible by computer analysis of the root maps. Statistical tests of spatial correlation between roots and other map features (e.g. cracks, tree roots, termite channels) can be performed on the basis of root intensity in zones with increasing distance to the features of interest (van Noordwijk *et al.*, 1992). The relation between root distribution and the pattern of cracks and macropores can be studied by infiltrating a methylene blue solution (0.05 g/l) into the soil before digging an observation pit.

Counts of N per horizon indicate the relative decrease of root intensity with depth. Root length density, L_{rv} (cm root length per cm^3 of soil) can be estimated for each horizon:

$$L_{rv} = 2 \times F \times N$$

where the calibration factor F is equal to 1.0 for roots with random orientation, if all roots are seen. The factor F depends on root orientation (between 0.5 and infinity) and improved

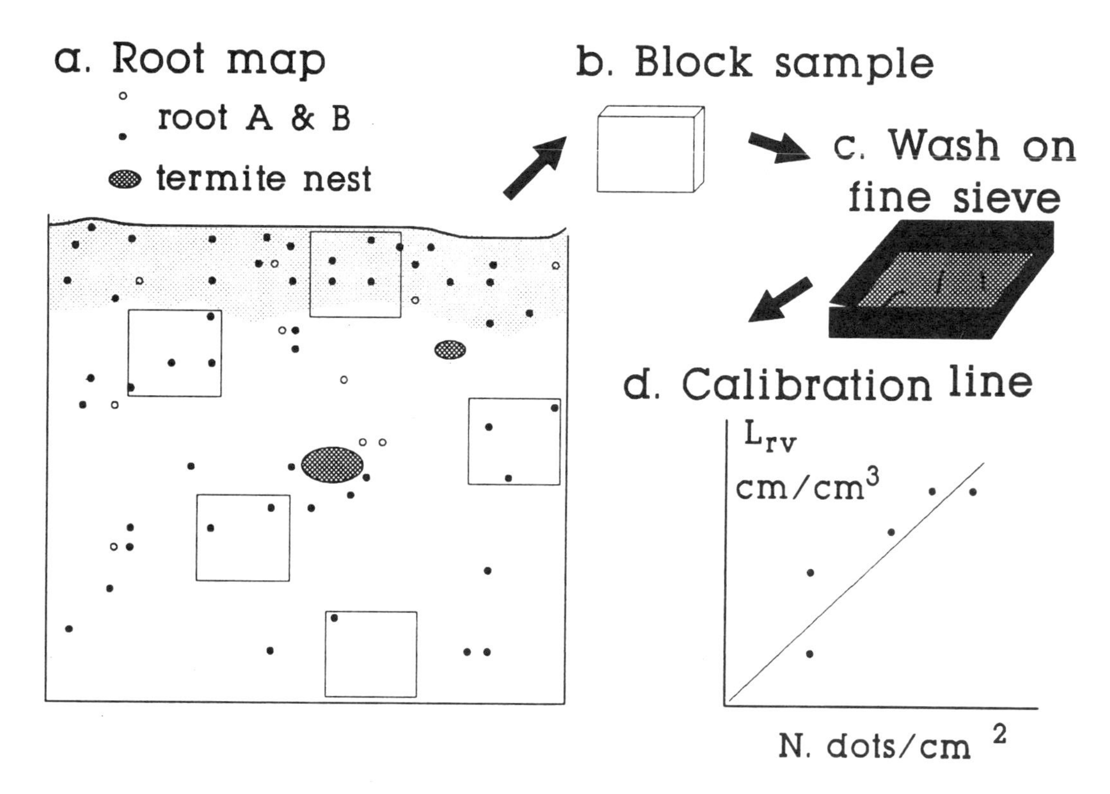

Figure 3.2 Root mapping on PVC sheets; roots A and B might refer to different diameter classes or species.

estimates can be obtained if root intensities in horizontally oriented planes are recorded as well (van Noordwijk, 1987). Experimental calibration factors may be considerably (e.g. three-fold) larger than theoretical ones.

Problems with the method are: (i) roots of different plants are hard to distinguish, (ii) distinction of live and dead roots is not easy and (iii) a considerable fraction of fine roots may be overlooked. To obtain (semi)quantitative results it is necessary to calibrate the maps by taking small blocks of soil (e.g. 20 x 10 x 1 cm volume) from various layers on the root map (Figures 3.2 b, c, d), wash them over a fine sieve (0.3 mm mesh) and determine root length (see below).

3.3.3 Pinboard monolith sampling

Monolith samples can be obtained with pinboards ("fakir beds"), made by inserting U-shaped pins (made from stainless steel) in plywood (Figure 3.3; further details are given by Schuurman and Goedewaagen, 1971, and Bohm, 1979). The size of the pinboard is determined by the crop (based on previous observations, such as rooting depth and distribution and practical considerations (samples of 100 x 60 x 10 cm of soil will weigh about 100 kg). By washing the soil away the roots become exposed and can be observed. If a coarse mesh screen (e.g. material used for TSBF litter bags) is put on the pins before the board is pushed into the soil (perpendicular to the crop row), this screen can help to keep the roots in their original location while washing the sample. Washing the sample can be facilitated by soaking overnight in water, deep freezing (for clay soils), soaking in oxalic acid (for soils with free calcium carbonate) or soaking in hexametaphosphate, preferably under vacuum. Whatever the pretreatment used, gentle washing must follow.

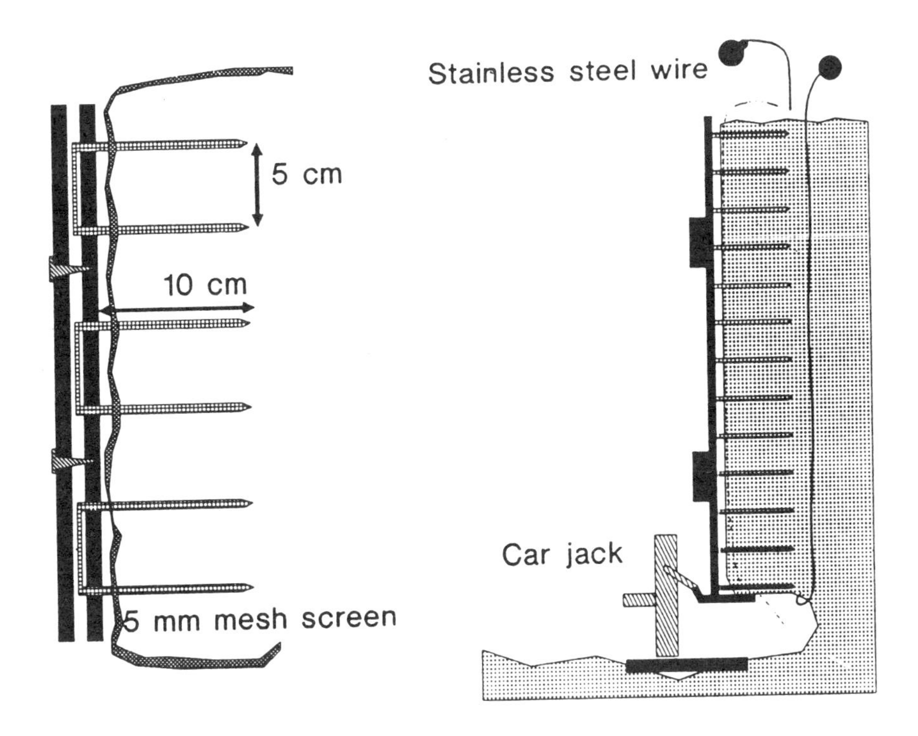

Figure 3.3 Pinboard construction and sampling in the field; after pushing the pinboard into a plane profile wall, the board is supported on a car jack and soil below and to the sides of the board is cut away; extending pieces of the mesh screen can be used to tie the soil block to the board; finally the back side of the sample is cut by "sawing" movements with a stainless steel wire.

After washing away the soil, the root system can be lifted on the mesh screen, photographed (on a black cloth as background) and/or cut according to soil layers (indicated e.g. by a string between the pins while washing the sample), depth zones and/or distance to the plant to obtain root biomass and/or root length (see below for root length). To estimate total biomass per plant root weight density per zone and depth has to be integrated over the relevant volume.

Although the pinboard method is more time consuming than methods 3.3.1 and 3.3.2 above, it gives good information per unit effort spent. Major weaknesses are that roots may break or be displaced during washing. The distinction between live and dead roots is easier via pinboard sampling then in methods where the root system is not sampled in its entirety.

References

Böhm, W. (1979) *Methods of Studying Root Systems.* Springer-Verlag, Berlin.

Caldwell, M.M., and Virginia, R.A. (1991) Root systems. In: Pearcy, R.W., Ehleringer, J., Mooney, H.A. and Rundel, P.W. (eds.), *Plant Physiological Ecology: Field Methods and Instrumentation.* Chapman and Hall, London, pp.367-398.

Floris, J. and van Noordwijk, M. (1984) Improved methods for the extraction of soil samples for root research. *Plant and Soil* **77**, 369-372.

Mackie-Dawson, L.A. and Atkinson, D. (1991) Methodology for the study of roots in field experiments and the interpretation of results. In: Atkinson, D. (ed.), *Plant Root Growth, an Ecological Perspective.* Blackwell Scientific Publications, Oxford, pp.25-47.

Schuurman, J.J. and Goedewaagen, M.A.J. (1971) *Methods for the Examination of Root Systems and Roots.* Second edition, Pudoc, Wageningen, The Netherlands, 86pp.

Taylor, H.M., Upchurch, D.R., Brown, J.M. and Rogers, H.H. (1991) Some methods of root investigations. In: Persson, H. and McMichael, B.L. (eds.), *Plant Roots and their Environment.* Elsevier Science Publishers, Amsterdam, pp.553-564.

van Noordwijk, M. (1987) Methods for quantification of root distribution pattern and root dynamics in the field. *Proceedings of the 20th Colloquium of the International Potash Institute*, Baden bie Wien, pp.243-262.

van Noordwijk, M., Widianto, Heinen, M. and Hairiah, K. (1991) Old tree root channels in acid soils in the humid tropics: important for crop root penetration, water infiltration and nitrogen management. *Plant and Soil* **134**: 37-44.

van Noordwijk, M. and de Willigen, P. (1991) Root functions in agricultural systems. In: Persson H. and McMichael, B.L. (eds.), *Plant Roots and their Environment.* Elsevier Science Publishers, Amsterdam, pp.381-395.

van Noordwijk, M., Brouwer, G. and Harmanny, K. (1992) Concepts and methods for studying interactions of roots and soil structure. *Geoderma* (in press).

3.4 LITTER DECOMPOSITION

3.4.1 Introduction

Decomposition is a complex process regulated by the interactions between organisms (fauna and micro-organisms), physical environmental factors (particularly temperature and moisture) and resource quality (defined here by lignin, nitrogen and condensed and soluble polyphenol concentrations) (Swift *et al.*, 1979). As decomposition progresses, soluble and particulate materials from the litter, organism tissues and products of microbial metabolism are separated from the original resource by leaching, physical fragmentation and animal feeding activities. These products are then transported by wind, water and gravity to soil microhabitats which have a different set of conditions regulating decomposition to those of the parent material. An almost intractable problem in quantifying litter decomposition is the need to impose methods which enable the experimental material to be identified without affecting the variables which regulate the component processes.

3.4.2 Litter bags and decomposition rate constants

Enclosing litter in a mesh bag makes it possible to recover the residual experimental material and defines the conditions under which the organisms operate. However, the mesh bag and compaction of the litter can create different microclimates to conditions in unconfined litter and the incorporation of material into soil is also affected. A modified method is described in Section 3.4.4 which overcomes some of these limitations (while imposing others). None the less the standard litter bag remains one of the most convenient methods for comparative measurements of decomposition between sites and treatments.

The minimum requirement for litter decomposition rates in TSBF studies is the time for 50% loss of the initial mass (t_{50}) but most studies in the literature report the decomposition rate constant "k" from the negative exponential function:

$$W_t / W_0 = W_0 e^{-kt}$$

where W_t is the amount of the initial mass (W_0) remaining at time "t".

This is more conveniently expressed as the regression function:

$$\log_n (W_t / W_0) = \log_n W_0 - kt$$

The use of the single exponential model assumes that decomposition rates are constant over time. In fact, litter mass losses often show a fast initial phase due to leaching of water soluble materials (which is largely an experimental artifact of drying and re-wetting the litter), a slower second phase dominated by the decomposition of cell wall constituents such as hemicellulose and cellulose, and a slow phase regulated by lignin and microbial products. This series of curves can be described by fitting a two- or three-phase exponential model with a rate-constant for each of the fractions (Wieder and Lang, 1982).

3.4.3 Decomposition of unconfined litter and turnover rates

Measurements of mass losses from unconfined litter under natural conditions are only possible in the initial phases of decomposition before the material starts to disintegrate. The method is most satisfactory for wood and other discrete materials which can be tagged or attached to a tether (Lang and Knight, 1979). For branches and tree boles changes in relative density (weight per unit volume) provides a useful measure of mass loss without sacrificing the whole unit. Extensive decomposition of wood often occurs before branches and tree boles fall to the ground. This can be determined by comparing the relative densities of standing dead and forest floor material with live wood but the age of the material is difficult to calibrate (Swift *et al.*, 1976).

Litter decomposition rates on the soil surface can be estimated from changes in litter standing crops over intervals of time where (i) there is a pulsed input of material (such as crop residues at harvest), (ii) the material has decomposed before the next input, or (iii) age classes of materials can be distinguished. In most natural systems it is often difficult to distinguish litter cohorts; especially where litter fall is more or less continuous or where litter is non-deciduous (e.g. grasses). Under these conditions determinations of litter turnover rates provides a useful basis for comparing litter dynamics in different habitats in relation to climate and site factors. The turnover coefficient "k" is calculated as the quotient of litter input (I) divided by the standing crop (S):

$k = I / S$

Even for small litter, the same fractions must be represented in the inputs and standing crops when calculating the quotient and this is usually only possible for leaf litter and small twigs (Anderson and Swift, 1983). Sites subject to fires (Woods and Raison, 1982) and heavier grazing present obvious difficulties in estimating litter turnover by input/standing crop determinations.

It is unfortunate that, for historical reasons, the turnover-coefficient and decomposition-rate constant, which have no formal relationship, are both designated by "k" and the use of this term must therefore be specified.

3.4.4 Litter bag methods

Modified method

For studies in agricultural systems, where inputs of crop residues, green manures or dung have largely decomposed by the start of the following cropping season, an alternative procedure has been developed using bags as mesh cylinders to retain the experimental material in contact with the ground. The method is most appropriate for surface litter treatments; sampling and processing buried litter requires modification of the protocol depending on the material and soil texture.

The advantages of the method are that it reduces microclimatic artifacts, allows natural access by soil fauna, enables measurements to be made of litter incorporation into soil and sampling can also be combined with determinations of other soil properties and processes. The disadvantages are that the method involves disturbing the soil (hence is most appropriate for arable systems) and samples take longer to process. The method is also less suitable where litter fall occurs during the experimental period though this can be excluded by fine mesh covers (e.g. mosquito-net). The method is unsuitable where there is a residual standing crop of litter.

Procedure

1 Open the bags to form mesh tubes (ca. 20 cm diameter by 30 cm deep) and bury them to about half of the depth in the soil. For stony or shallow soils the bag can be cut in half and only 5 cm inserted below ground. Placement is made easier by using a metal sleeve which fits inside the bag to hold the cylinder rigid during handling. In tilled soils, large stones may be removed from within the cylinder to facilitate subsequent sampling.

2 Locate the mesh cylinders on a random basis in the field and add a known mass of experimental material in parallel with plot treatments.

3 Sample 3 to 5 bags on a random basis at regular intervals. Collect as much of the litter from the soil surface as possible (this may include small stones and aggregates) and put into a bag.

4 *Surface-litter treatments*: soil cores (ca. 5 cm diameter) can be taken after the removal of litter and divided into depth increments to determine moisture content, organic carbon distribution, soil organic matter fractions and N-mineralization rates in the laboratory. The remaining material can be dug out and hand sorted for roots and soil macro-fauna.

Buried-litter treatments: take cores for associated measurements and then remove all the soil in the bag to a suitable location for processing.)

5 Fill in the hole with soil from the margin of the plot and mark the location to avoid resampling.

6 *Surface-litter treatments*: in the laboratory sort the bagged litter material by hand, or sift in a coarse sieve, to separate the large litter fragments, stones and soil aggregates such as termite/earthworm worked material.

Place the residual litter and soil material in a bucket of water, briefly swirl to suspend litter fragments, and carefully decant through a 2 mm mesh sieve. Experimental material retained on the sieve is classified as litter. Use a 0.25 mm sieve to further separate material in the washings if soil litter is to be determined (see Section 6.5.5).

Buried litter treatments: hand-sort the material for litter (fauna and roots) as far as practical. Subsample the soil and separate by water flotation as described above. In soils with high clay content, subsamples may be deflocculated with a water softener to release occluded material.

7 Drain washed litter thoroughly before air drying. Take subsamples for oven dry weight determinations.

Note: Little leaching occurs if washing is brief but this can be readily checked. It may be impractical to use distilled water for this separation but the water chemistry should be borne in mind in subsequent analyses. Litter which is heavily contaminated by clay material may be separated by more prolonged washing or sonication but effects on nutrient leaching must be assessed. Water softeners used in defloculation may have high concentrations of phosphorus, so samples treated that way may not be suitable for nutrient analyses. In some cases, such as where termites have infilled woody materials with soil, it may be necessary to ash the experimental material and express results on an ash-free basis.

Data analysis

Plot the data for % initial mass, and $\log_n$% initial mass, vs. time (days) to inspect for trends.

If the untransformed data show linear trends, estimates of t_{50} are made by calculating the regression and solving the equation for t = 50%.

The convention is to fit a single negative exponential function to derive the decomposition rate-constant "k"; t_{50} can then be estimated from the reciprocal of k/day.

The rate-constant "k" is the slope of the regression of $\log_n$% remaining vs time (Equation 2, Section 3.4.2). If time is in days then the slope is -$\log_n$%/day. To obtain k/year multiply by 365.

For example, for *Shorea affinis* (sampled over 294 days in a rainforest in Sri Lanka), k = -0.0036/day; k/year = -0.0036 x 365 = 1.22/year (values for k are conventionally expressed as positive values). The estimate of t_{50} = 1/0.0036 = 278 days.

3.4.5 Decomposition standards

Birch (*Betula spp.*) lollipop sticks are used as a decomposition standard for inter-site and regional comparisons using natural vegetation (where possible) and a derived site in each study area.

The sticks (100 x 10 x 1 mm) are secured in bundles of three with a wire or plastic tie at each end of the bundle. Two sets of bundles are used to investigate the effects of litter and soil conditions on decomposition rates.

Procedure

1 Place one set horizontally on the soil surface and bury the other set upright in the soil with the tips at the soil surface. Pairs of bundles, one of each set, are placed and samples on a random basis.

2 Collect samples of 10 bundles of each set at intervals of 3 months.

3 Remove the middle stick from the bundle, dry at 85°C and weigh.

A parallel series of sticks enclosed in fine stainless steel mesh (0.5 mm) is also employed at TSBF sites to assess termite effects on mass loss.

References

Anderson, J.M. and Swift, M.J. (1983) Decomposition in tropical forests. In: Sutton, S.L., Whitmore, T.C. and Chadwick, A.C. (eds.), *Tropical Rain Forest: Ecology and Management*. Blackwell Scientific Publications, Oxford, pp.287-309.

Lang, G.E. and Knight, D.H. (1979) Decay site for boles of tropical trees in Panama. *Biotropica* **11**, 316-317.

Swift, M.J., Healey, I.M., Hibberd, J.K., Sykes, J.M., Bampoe, V. and Mesbitt, M.E. (1976) The decomposition of branch-wood in the canopy and floor of a mixed deciduous woodland. *Oecologia* (Berl.) **26**, 139-149.

Swift, M.J., Heal, O.W. and Anderson, J.M. (1979) *Decomposition in Terrestrial Ecosystems*. Blackwell Scientific Publications, Oxford.

Wieder, R.K. and Lang, G.E. (1982) A critique of the analytical methods used in examining decomposition data obtained from litter bags. *Ecology* **63**, 1636-1642.

Woods, P.U. and Raison, R.J. (1982) An appraisal of techniques for the study of litter decomposition in eucalypt forests. *Australian Journal of Ecology* **7**, 215-225.

3.5 SOIL

3.5.1 Soil CO_2 evolution

The efflux of CO_2 from soil ("soil respiration") theoretically represents an integrated measure of root respiration, soil fauna respiration and the carbon mineralisation from all the different carbon pools in soil and litter. Over the year, the total efflux should correspond to above- and below-grown inputs from plants under steady-state conditions. Latitudinal gradients of soil respiration in natural systems broadly follow trends in above- and below-ground litter production (Raich and Nadelhoffer, 1989), but the additional component of CO_2 production derived from root respiration can constitute up to 50 % of the flux (Singh and Gupta, 1977; Coleman and Sasoon, 1980; Raich and Nadelhoffer, 1989). Precise allocation of CO_2 production to plant or soil processes is therefore technically difficult and generally involves the use of ^{14}C tracer techniques. Measurements of CO_2 evolution do, however, provide a sensitive indication of the response of microbial activity to diurnal variations in temperature and moisture, the effects of wetting and drying following rainfall events, and the differences in plot treatments (mulching, tillage, irrigation, etc.).

The CO_2 produced at the soil surface is collected in airtight chambers placed over the soil (including the litter layer if present) and trapped in an absorbent. The specifications of the chambers are not precise, and their size can be varied as appropriate to the plot design (the use of the modified plastic bucket is described below). It is however important for short-term measurements, or when rates of CO_2 afflux are low, to keep the enclosed volume of air as small as possible and determine the CO_2 content of the head-space. Measurements of CO_2 production from large areas over long periods of time avoid these problems but obviously integrate across short-term variations in microbial activity which may be of interest. The presence of the chambers can also modify soil temperature and moisture conditions, and hence rates of microbial processes. It is therefore essential to check on similarities of soil microclimates inside and outside the chambers.

Procedure

1. Cut hard plastic buckets (of about 27 cm diameter at the base) around the circumference about half way down its height; (Figure 3.4 a). The bucket bottoms will make a set of chambers about 20 cm high, of approximately 560 cm^2 area at the open end, while the bucket tops form a set of cylinders. When inverted (open end down), the bucket bottoms will fit within the cylinders to make an airtight seal. Drill a 2 cm hole in the bottom of the bucket and fit it with a rubber stopper.

2. One week prior to measurement, insert the plastic cylinders (bucket tops, cut edge down) no more than 3 cm into the soil surface at random locations within the study plot; (Figure 3.4 b). Cut and remove all living vegetation within the cylinders at the soil surface taking care to minimise soil and litter disturbance.

3. One week later place an open tin can containing soda lime (see below) in the centre of each cylinder and immediately fit the chamber tightly into the cylinder, having first removed the stopper. Replace the stopper once the chamber is firmly jammed in the cylinder.

4 After 24 hours remove the chambers, close the tin cans and remove them, and remove the cylinders from the soil.

Note: Never place chambers in the same location twice, or leave the cylinders in place from one measurement period to another.

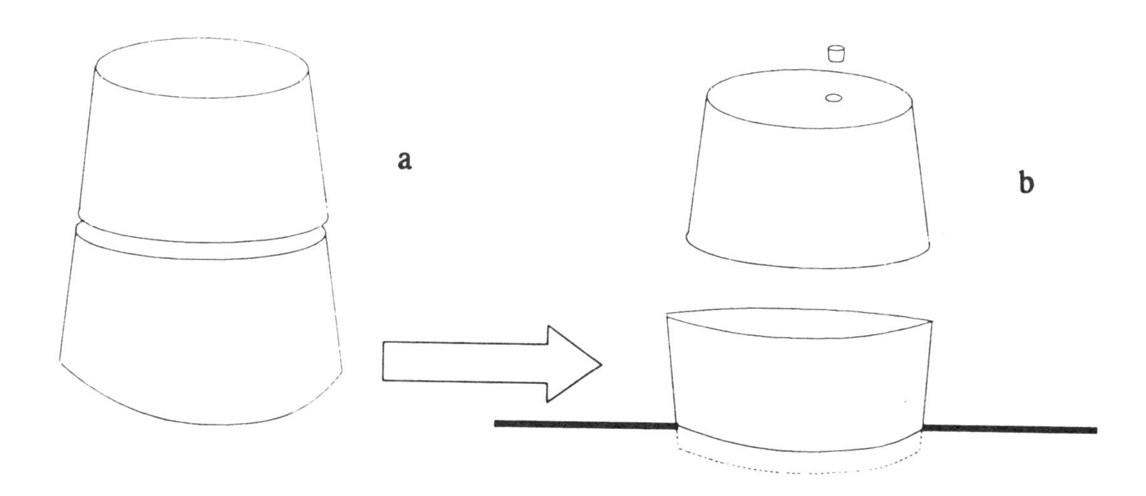

Figure 3.4 Using a modified plastic bucket as a collector for soil-respired CO_2.

Making and using the soda lime traps

Use about 40 g of reagent-grade soda lime (6-12 mesh) in a 5.2 cm tall, 8.1 cm diameter soil tin with a tightly fitting lid. Such tins will cover about 9% of the surface area of the floor enclosed by the chamber for the bucket size above. The 40 g of soda lime follows the recommendation of Edwards (1982) of using a minimum amount of 0.6 g/cm^2 of floor; the minimum amount of soda lime for a chamber with an area of 560 cm^2 would be 34 g.

Procedure

1 Weigh an empty tin can and lid (W_t). Add about 40 g of soda lime (or the amount required according to chamber size), to each tin and dry at 105°C for 24 hours. Remove the tin can from the oven, put the lid tightly on and allow to cool. When cool, weigh to the nearest 0.01 g (W_b). (Use blank tins to account for CO_2 absorption during handling and oven drying; these are similarly treated tin cans filled with soda lime except that they remain closed during the measurement period.)

2 Take the tin cans to the field, uncover, place within the cylinder making sure that the lid is under the tin can and place the chamber tightly. Leave in the field for 24 hours.

3 After 24 hours remove the tin can from the chamber, put the lid tightly on, and take to the laboratory for drying at 105°C for 24 hours. Remove the tin can from the oven, put the lid on and allow to cool. When cool weigh to the nearest 0.01 g (W_a). Make sure that the can is closed tightly.

4 Calculate the amount of CO_2 evolved from the soil from the mass gain of the soda lime sample minus the average blank gain. (The factor 1.4 compensates for chemical water lost during the drying process.)

Calculations

CO_2 evolved/chamber (g) = $[(SL_a - SL_b) - B] \times 1.4$

CO_2 evolved/m^2/hr (mg) = $1000 \times \{(SL_a - SL_b) - B\} \times 1.4 \times \{10000 / (A \times t)\}$

where
SL_a (g) = weight soda lime after field placement = $W_a - W_t$
SL_b (g) = weight soda lime before field placement = $W_b - W_t$
B (g) = mean blank soda lime mass gain
A (cm^2) = chamber area
t (hr) = exposure time

Note: Use soda lime no more than twice if CO_2 uptake is less that 5% of initial weight of soda lime. If weight gain is greater than 2 g do not reuse the soda lime.

Note: For short-term measurements a volumetric procedure can be more sensitive than the gravimetric procedure described above. This uses aliquots of 0.1 M KOH exposed in plastic cups for 1 - 2 hr followed by titration with standardised 0.1 M HCl after the addition of saturated $BaCl_2$ solution (2 ml per 25 ml KOH). $BaCl_2$ is poisonous - do not pipette by mouth. Larger volumes of absorbent or molar solutions can be used if 0.1 M KOH becomes saturated during the exposure period. Absorbed CO_2 is calculated on the basis that 1 ml 0.1 M HCl is equivalent to 2.2 mg CO_2; or 1 ml 1.0 M HCl is equivalent to 22 mg CO_2.

References

Anderson, J.P.E. (1982) Soil Respiration. In: Page, A.L., Miller, R.H. and Keeney, D.R. (eds.), *Methods of Soil Analysis, Part 2. Chemical and Microbiological Properties. Agronomy Monograph* **No. 9.** (2nd edition), ASA, Madison, pp.831-871.

Coleman, D.C. and Sasoon, A. (1980) Decomposer sub-system. In: Breymayer, A.I. and van Dyne, G.M. (eds.), *Grasslands, System analysis and Management.* Cambridge University Press, Cambridge, pp.609-655.

Edwards, N.T. (1982) The use of soda-lime for measuring respiration rates in terrestrial systems. *Pedobiologia* **23**, 321-330.

Raich, J.W. and Nadelhoffer, K.J. (1989) Below-ground carbon allocation in forest ecosystems: global trends. *Ecology* **70**, 1346-1354.

Singh, J.S. and Gupta, S.R. (1977) Plant decomposition and soil respiration in terrestrial ecosystems. *Botanical Review* **43**, 449-475.

3.5.2 Soil fauna

Functional classification

Soil invertebrates can be classified according to their feeding habits and distribution in the soil profile as follows:

1. *Epigeic* species which live and feed on the soil surface. These invertebrates effect litter comminution and nutrient release but do not actively redistribute plant materials (though the comminuted material may be more easily transported by wind or water than the material from which it was derived). These species are mainly arthropods (e.g. Myriapods, Isopods or fairly small and entirely pigmented earthworms).

2. *Anecic* species which remove litter from the soil surface through their feeding activities (e.g. earthworms dragging litter into their burrows or termites removing litter to mounds). Considerable quantities of soil, mineral elements and organic matter may be redistributed through these activities accompanied by physical effects on soil structure and hydraulic characteristics.

3. *Endogeic* species which live in the soil and feed on organic matter and dead roots. The two main groups are earthworms and humivorous termites which can turn over a significant proportion of top soil and soil organic matter per year in some sites.

The quantification of these effects on soil processes requires a detailed study but a simple characterisation of the macrofauna is required to assess their role in different systems and the impact of management practices on population densities and community structure. See Table 3.1 for a list of generalised taxonomic units and functional classification.

Sampling

The recommendation is for a minimum of 5 (but preferably 10) 25 cm x 25 cm by 30 cm deep soil monoliths sampled towards the end of the rainy season and hand-sorted for macro-invertebrates (body length < 2 mm). It must be stressed this technique is designed for undertaking a soil fauna survey; manipulative experiments may need other sampling strategies. Several references related to sampling macrofauna are given at the end of this section.

Procedure

Locate the sampling points 5 m apart along a transect with a random origin.

1. Remove litter from within a 25 cm quadrat and retain for sorting.

2. Isolate the monolith by cutting down with a spade a few centimetres outside the quadrat and then digging a 20 cm wide by 30 cm deep trench around it. (This facilitates cutting of the sample into horizontal strata and collecting animals escaping from the block.)

Table 3.1 Taxonomic units for soil fauna site characterisation and their functional classification (ecological category).

Taxonomic units	*Ecological category*
Ants	Epigeic/Anecic
Arachnids (spiders, etc)	Epigeic
Beetle adults	Epigeic/Endogeic
Beetle larvae	Epigeic
Blattoidea (cockroaches)	Epigeic
Chilopoda (centipedes)	Epigeic
Cicadidae	Endogeic
Diplopoda (millipedes)	Epigeic
Earthworms, pigmented	Epigeic/Anecic
Earthworms, un-pigmented	Endogeic
Gastropods (slugs and snails)	Epigeic
Gryllidae (crickets)	Epigeic
Isopoda (woodlice)	Epigeic
Termites	Anecic/Endogeic
Other groups	Various

3 Collect all invertebrates longer than 10 cm excavated from the trench; these will mainly be large millipedes and earthworms with very low population densities but representing an important biomass. (Their abundance and biomass per m^2 of these groups is calculated on the basis of 0.42 m^2 samples, i.e. the width of the block plus two trench widths, squared.)

4 Divide the delimited block into three layers, 0 - 10 cm, 10 - 20 cm, and 20 - 30 cm, which are then hand-sorted.

Note: In sites where soil fauna have low abundance and/or the above method produces unacceptable disturbance:

Remove the 0 - 10 cm soil layer intact and then dig out the remaining sample to a 30 cm depth with a spade. Cut the sides of the sample with the spade as rapidly as possible (in advance of digging out) to isolate the block and limit the loss of termites from the sample (which is counted as a single unit).

5 Hand-sort soil and litter material in trays about 50 cm x 30 cm x 5 cm deep: deposit a handful of soil on the left side of the tray and progressively move it towards the right side, over the whole surface. After removal of the animals from the tray, empty it, and process another sample. This method achieves a high sorting efficiency, and is faster than having large quantities of material in the tray to process at a time.

6 When possible, separate further separate subsamples by wet sieving to assess the accuracy of the method.

7 Preserve the invertebrates in 4% formaldehyde, and keep earthworms separate from other groups.

8 Record numbers and fresh (preserved) weight of invertebrates for the litter and other strata according to the taxonomic units listed in Table 3.1. Finer taxonomic subdivisions are desirable if the expertise is available. Termite fungus combs should also be recorded.

Note: Large aggregations of ants or termites in woody litter, or soil chambers exposed on the sides of the sample, are best collected as bulk material and sorted in the laboratory.

In sites where termite mounds or ant nests are a conspicuous feature enumerate the mounds in a suitable area and calculate densities on a hectare basis.

Further reading

Bouche, M.B and Gardener, R. (1984) Earthworm functions VII. Population estimation techniques. *Revue d'Ecologie et de Biologie du Sol* **21**, 37-63.

Darlington, J.P.E.C. (1984) A method for sampling the populations of large termite nests. *Annals of Applied Biology* **104**, 427-436.

Ernsting, G. (1988) A method to manipulate population densities of arthropods in woodland litter layers. *Pedobiologia* **32**, 1-6.

Grace, J.K. (1990) Mark-recapture studies with *Reticulitermes flavipes* (Isoptera: Rhinotermitidae). *Sociobiology* **16**, 297-303.

Haverty, M.I., Nutting, W.L. and Lefage, J.P. (1976) A comparison of two techniques for determining abundance of subterranean termites in an Arizona desert grassland. *Insectes Sociaux* **23**, 175-178.

Kretzshmar, A. (1978) Ecological quantification of burrow systems of earthworms. *Pedobiologia* **18**, 31-38.

Lavelle, P. (1988) Assessing the abundance and role of soil macroinvertebrates communities in tropical soils: Aims and methods. In: Ghabbour, S.I. & Davis, R.C. (eds.), Proceedings of the Seminar on Resources of Soil Fauna in Egypt and Africa, Cairo, 16-17 April 1986. *African Journal of Soil Zoology* **102**, 275-283.

Lavelle, P. and Kohlmann, B. (1984) Étude quantitative de la macrofaune du sol dans une forêt tropicale humide de Mexique (Bonampak, Chiapas), *Pedobiologia* **27**, 377-393.

MacCauley, B.J. (1975) Biodegradation of leaf litter in *Eucalyptus pauciflora* communities I. Techniques for comparing the effects of fungi and insects. *Soil Biology and Biochemistry* **7**, 341-344.

Springett, J.A. (1981) A new method for extracting earthworms from soil cores, with a comparison of four commonly used methods for estimating earthworm populations. *Pedobiologia* **21**, 217-222.

Chapter 4 SAMPLING FOR LABORATORY ANALYSIS AND SAMPLE PREPARATION

4.1 SOIL AND PLANT SAMPLING FOR LABORATORY ANALYSIS

Sampling and sample preparation is arguably the most important stage of any study; excellent laboratory analysis does not compensate for poorly collected, poorly prepared or unrepresentative samples. Whenever possible, label sample containers prior to going to the field. Field-fresh soil samples are often moist; these should be collected in plastic bags or tubs rather than paper bags. Plant materials, on the other hand, should be collected in paper bags to reduce decomposition accelerated by a moist environment prior to preparation for analysis. Upon return to the laboratory, examine plant samples for contamination, e.g. dust or mud; a brief rinse with water may be required.

Unless the analysis requires a field moist sample, air-dry all samples as soon as possible to halt biological transformations.

4.1.1 Soil sampling

Augering procedure

1 Remove surface plant and litter material from a minimum of 5 randomly selected auger sites.

2 At each site auger to a known depth (e.g. 30 cm) and put the auger contents in a clean plastic bucket so that all 5 augerings can be thoroughly mixed prior to subsampling. Take a subsample at about 500 g (one cupfull). Discard the remaining soil.

3 Repeat step 2 for the next depth required for each auger hole.

Undisturbed cores procedure

1 Clean a flat soil surface (either a vertical face in a profile pit or a horizontal surface at any known depth).

2 Firmly drive a soil coring cylinder (of known volume and mass) fully into the face causing minimum disturbance and compaction to the core.

3 Cut away surrounding soil so as to be able to cut neatly across the inserted end of the cylinder.

4 Carefully lift the full core away, trim the ends flush with the cylinder ends, and cap tightly.

4.1.2 Plant sampling

The part of the plant that needs to be sampled and the sampling strategy depends on (i) the purpose to which the samples will be put (i.e. the types of analyses), and (ii) the type of plant to be sampled. No standard method is thus applicable. The general concept as discussed for soils applies, *vis.* taking several (minimum 5) subsampless, mixing them to make a composite sample and finally taking a subsamples for analysis. Sampling for litter and crop residues is described in Section 3.2 above.

4.2 SOIL, PLANT AND LITTER MATERIAL PREPARATION

4.2.1 Soil material

Note: As many of the variables of interest in TSBF studies are biotic, or biologically mediated, several of the analyses must be conducted on field-fresh soil samples; the following guidelines relate to soils that do not need to be field-fresh.

Procedure

1 Air-dry the soil by spreading it out in a shallow tray in a well ventilated place protected from rain and contamination. Alternatively soils can be dried in a forced air oven at a maximum temperature of 60°C. Break up any clay clods. When the soil is dusty it is dry enough.

2 Crush the soil lumps gently so that the gravel and roots etc. are separate from the mineral soil.

3 Sieve the soil through a 2 mm sieve leaving the gravel and roots etc. in the sieve.

4 Pick out the roots and save if required.

5 Retain the gravel for weighing if required. This should be done if it appears to be about 5% or more by mass of the original sample.

6 Retain a representative sample (of the sieved soil approximately 250 g), by e.g. coning and quartering, for analysis.

Note: Coning and quartering is a simple and effective method of subsampling loose, < 2 mm material: mix, then pile the material in a flat cone or mound. Divide into four quarters by cutting across the middle in an "X" shape with a blade. Discard two opposite segments, and mix and re-cone the other two. Repeat the process until sufficient material remains.

4.2.2 Plant material and litter

Procedure

1. Dry green material and litter at a maximum of 60°C; dung should be dried at a maximum of 40°C to prevent volatilisation.

2. Grind all the material to pass a 0.15 mm mesh.

3. Retain a representative sample of approximately 10 g, by e.g. coning and quartering (see Section 4.2.1), for analysis.

4.2.3 Sample storage

Both soil and plant material will satisfactorily store for long periods (i.e. > 1 year) if they have been correctly prepared. Samples are stored in clearly labelled, air-tight, termite-proof containers; the storage room should be well ventilated.

Chapter 5 LABORATORY PRACTICE AND QUALITY CONTROL

5.1 LABORATORY LAY-OUT

The main aim in designing the lay-out of the laboratory is to create a work environment which is ergonomically efficient. If this is achieved, many of the issues discussed below, especially safety, throughput and reproducibility, are maximised. In general, samples should "flow" smoothly through the lab, (i.e. without being repeatedly carried backwards and forwards), moving away from the reception and preparation areas to cleaner areas for analysis. The designated work areas (see Table 5.1) should be chosen to afford maximum ease of operation, and a level of cleanliness required for the task in hand. As few labs are ideal in design and location, compromise inevitably has to be sought.

5.2 SAFETY

Many laboratory procedures involve handling hazardous substances and eye protection and laboratory coats must be used. Supplies of other protective equipment (e.g. thick rubber gloves for cleaning acid spills, tongs etc.) should be provided. An eye wash station should be readily accessible near the area of maximum hazard, probably the fume cupboard, and all lab personnel must be familiar with its use. Regularly inspected CO_2 or foam fire extinguishers should be easily accessible.

Only those reagents and items of equipment needed for a given procedure should be in the work area in question; back-up supplies should be kept in a store. All reagent bottles must be clearly marked, including any particular hazard information, e.g. poison or strong acid. Stores and cabinets for hazardous chemicals should be secure and lockable. Flammable chemicals should be kept in separate, flameproof cabinets. Procedures involving strong acids and/or high temperatures must be performed in an efficient fume cupboard.

Any spills of chemicals, solutions or water should be promptly cleared up, and all waste bins regularly emptied. Work areas should be kept "clutter-free", and a good supply of cleaning equipment, cloths and tissues should be provided.

Liquid waste should be poured carefully down a sink, with plenty of water to dilute and flush it away. A dilution tank should be installed between the lab's sinks and the main disposal site. Soil and plant analysis labs do produce toxic and acidic waste, but of a highly diluted form.

Soil extracts should be disposed of into a container, as disposal down a sink leads to severe plumbing maintenance problems. Eating, drinking and smoking must be prohibited in the lab at all times. All laboratory lab personnel must wash their hands with soap and water after lab work, and again before eating.

Table 5.1 Recommended work areas and tasks for standard soil and plant analysis.

Location	Task	Material when task completed
Field	Sample collection and labelling	Field condition soil and plant
Sample Reception	Sample bag sorting and ordering; label checks; recording entries in sample accession book	Field condition soil and plant
Soil 1° Sample Preparation Area	Air drying; crushing; 2 mm sieving; subsampling	2 mm air-dry soil
Plant 1° Sample Preparation Area	Oven drying; subsampling	Oven-dry plant
Soil 2° Sample Preparation Area	Subsampling; fine grinding	Finely-ground, air-dry soil
Plant 2° Sample Preparation Area	Subsampling; fine grinding	Finely-ground, oven-dry plant
Sample Store	*Soil:* 2 mm, air-dry *Plant:* ground, oven-dry Storing samples systematically with easy access and retrieval	*Soil:* 2 mm air-dry *Plant:* finely ground, oven-dry
Wet Chemistry Lab	Extraction; [digestion]; filtration; dilution; colour development; titration	Solutions
Instrument Lab	Determination (pH, EC, flame photometry, AAS, colorimetry); calculation	Instrument readings
Office	Calculation; report writing	Report, etc.

Note: Several soil analyses (e.g. mineral nitrogen extraction) will require field condition soils.

5.3 SERVICES

5.3.1 Water supply and quality

Soil and plant analysis labs require large and reliable supplies of water. Mains water is generally of adequate quality; the higher the quality, the better, as saline or dirty water rapidly shortens the lives of deionisers and water stills. If the mains supply is unreliable, a header tank should be installed, dedicated to the lab; it should not serve toilets and other general areas in the building.

Single distilled or deionised water should be used for all analytical work and reagent preparation, and the final stages of washing labware. Deionised water should have a conductivity value < 0.2 μS/cm.

5.3.2 Electricity supply and quality

Almost all analyses require electricity. The reliability of the supply is therefore of paramount importance. Electricity surges, lows and blackouts are an ever present threat in all labs, although the degree of risk may vary! All instruments should be surge-protected at least, to protect the circuits, and a stabilised electricity supply is essential for most. As most instruments are of low wattage, an uninterruptible power supply (UPS - as now widely used for computers) provides a very stable supply, and allows the completion of a set of sample readings in the event of a power cut; this last point is especially useful, as the instrument would have to be re-calibrated upon start up. In some locations (i.e. very humid or dusty), the most sensitive instruments will need to be in air-conditioned rooms with controlled humidity.

5.3.3 Work surfaces

Although Formica or plastic work surfaces are desirable, many labs have wooden benches. These are perfectly adequate providing they are kept clean and varnished. Disposable, plasticised sheeting can be obtained as an added precaution against contamination. The fume cupboard floor should be of specialised corrosive and heat resistant material, or glazed tiles.

5.4 GLASS AND PLASTICWARE WASHING

All glassware used in procedures must be scrupulously clean. Writing should be wiped off with a little acetone on a tissue; never use scouring powder, as it contains large amounts of phosphorus and chlorine, and scratches the article being "cleaned". Brushing in mains water, followed by at least two distilled or deionised water rinses will suffice in most cases. Detergent (phosphorus-free) need only be used if oily or greasy materials have been involved. More powerful cleaning agents such as chromic acid may occasionally be required.

Dirty plasticware is more difficult to clean because the internal and external surfaces are easily scratched. Writing may be removed by a mild solvent such as aqueous alcohol, as some plasticware is attacked by acetone. Wiping with a moist tissue, and soaking in phosphorus-free detergent solution or dilute acid, prior to thorough rinsing, will help to remove residues. When they become permanently marked or scratched the items should be discarded.

5.5 CONTAMINATION

Sources of contamination must be identified and eliminated. Some of the more common examples include:

- external dust blown in from the surrounding environment - try to minimise by closing windows in windy conditions;
- internal dust - this arises from cleaning operations, rusty fittings, plaster and decorating materials;
- inter-sample - only have one sample container open at once;
- reagents - store volatile reagents (especially ammonia) well away from samples;
- washing materials, particularly soap and scouring powders, and cosmetics.

As far as possible, keep a set of glassware exclusively for each type of analysis; for example, sets of bottles and test-tubes marked "P" should be used only for extractions and determinations of phosphorus.

5.6 LEVELS OF ACCURACY

A major saving in time can be achieved by deciding on the level of accuracy needed for a given operation. For instance, when making standards, maximum accuracy is needed in weighing out the reference salt, e.g. to make a 1000 ppm NH_4 solution, you must weigh out exactly 4.714 g of analytical grade ammonium sulphate, and make up to 1000 ml in volumetric glassware. On the other hand, when weighing out soil for a digest, maximum accuracy is needed to the required number of decimal places, although the actual amount is not critical, e.g. 1.089 g would satisfy the requirement of "about 1 g accurately", providing that the exact weight of sample is used in calculating the results.

The level of accuracy required for a given measurement can be inferred from the way it is stated in the instructions. Materials should be measured to the number of significant figures quoted; i.e. if a liquid addition is given as 4.4 ml, it should be greater than 4.35 ml and up to or including 4.44 ml.

Some reagents, generally those needed in excess, can be made to a low level of accuracy, and are often stated as "%" solutions. A 5% potassium dichromate solution need not be made volumetrically; 50 ± 5 g of GPR grade reagent dissolved (and mixed with a magnetic stirrer) in 1 litre from a measuring cylinder is sufficient.

5.7 SAMPLE BATCHES

When large numbers of samples have to be analysed, it is best to group them in batches. The batch size will usually be set by some aspect of equipment capacity - it may be the number of samples that can be simultaneously digested; or the number that can be completed, or brought to an appropriate pause in the procedure, within a working day.

In principle, the bigger the batch size, the better, providing the time to work from 1 to *n* is not so long as to detrimentally modify some aspect of 1 (e.g. colour fading); and providing that handling the large batch is convenient and the delay imposed at each stage of the procedure does not affect the results.

The greater the batch size, however, the greater the advantages of colorimetric over titrimetric determinations, as once calibrated, the colorimeter can be used more quickly (and often more accurately) than titrating each unknown.

Batches must include blanks, repeats and a reference sample (see Quality Control below). If a digestion system can accommodate 48 tubes, the batch would comprise new 40 samples, 4 repeats, 2 blanks, 1 reference and 1 aliquot of digest mixture for standard compensation.

5.8 DATA RECORDING

Data must be recorded in dedicated lab record books, and should include all data pertaining to a given analysis (e.g. date, name of analyst, sample identity, weight, titre or absorbance reading, etc.). The "raw data" books should never leave the lab. Data must be recorded directly into the book, and not on bits of paper, backs of hands, cigarette packets, etc! Calibration curve data should be recorded with the sample readings to which they pertain. Data should be recorded to the appropriate number of significant figures.

5.9 GENERAL POINTS

After calibrating the instrument with standards, sample readings must be done without further altering the instrument settings. If the power fails the instrument must be re-calibrated (see above under Electrical supplies).

Standards must contain equivalent levels of background components (e.g. acids or catalysts) as samples.

Standards and samples must be analysed as one batch with the same set of reagents.

Low level standards can be unstable and should be freshly prepared each day.

Reagents which are prone to oxidation must be made on the day of use - aim to make sufficient, but not excessive, quantities.

Sample solutions outside the analytical range quoted should be diluted bearing in mind they must be restored to the original reagent concentrations - never dilute the developed colour.

Dispensers and diluters must be thoroughly washed through with distilled or de-ionized water immediately after use, or the plungers may seize. To avoid contamination from metal parts or plastic seals, dispensers and diluters should be operated about 5 times with reagent, discarding the dispensings, before use.

5.10 QUALITY CONTROL AND STANDARDISATION PROCEDURES

Quality control is a very important aspect of lab practice. This is especially so in routine analysis labs where gradual drift can occur as a result of many factors, e.g. contamination, or changed reagents and environment, operator error etc. The aim is to achieve maximum reproducibility while maintaining adequate accuracy. A sample re-analysed after an interval should give the same result, within the accepted variation, as when initially analysed (except where for instance field condition soils are analysed).

Errors can be either consistent or random, depending on the causative factor. If observed, consistent errors are generally easier to rectify than random. There are several ways of monitoring quality.

5.10.1 Blanks

Blanks are reaction vessels that are subjected to identical procedures as the samples in a given batch, but have no sample added. They allow for corrections for any background levels introduced from reagents, filter papers, etc. Provided the blank values are the same, the mean value can be subtracted from the sample result. In most procedures at least two blank determinations should be included in a given batch. A very high blank value suggests contamination in the reagents, filter papers, etc. If observed, check, then repeat batch analysis.

5.10.2 Repeats within a batch

About 1 in 10 samples, selected at random from the batch, should be analysed in duplicate. (The choice of 1 in 10 is a compromise between the ideal of analysing all samples in duplicate, and time, effort and expense.) Obviously the answers for given pairs of duplicates should be the same (i.e. within ± 2 - 5% of the mean depending on the analysis in question), and any discrepancy must be investigated. Once rectified, the batch should be re-analysed.

5.10.3 Internal references

Internal reference materials are required for each type of material under investigation. Prepare a sufficient quantity of homogenous material for use in the initial testing of methods, training of technicians and the assessment of bias in subsequent analyses. Before commencing routine analyses obtain estimates of precision for use in quality control.

A sample taken from the well homogenised, bulk quantity should be included in each batch as an internal reference. If there was no analytical variation between batches, the values obtained for this sample would be the same in each batch. The variation from the mean value for this sample, calculated over previous batches, indicates the "batch error".

Plot each value of the internal reference on a quality control chart (e.g. Figure 5.1) to monitor the performance of the analysis. The "y" axis is the variable value, and the "x" axis is successive analyses of the internal reference.

Take action should a value exceed the ± 3 standard deviation (sd) limits, or if two successive values exceed the ± 2 sd limits. As more data are accumulated reassess the limits.

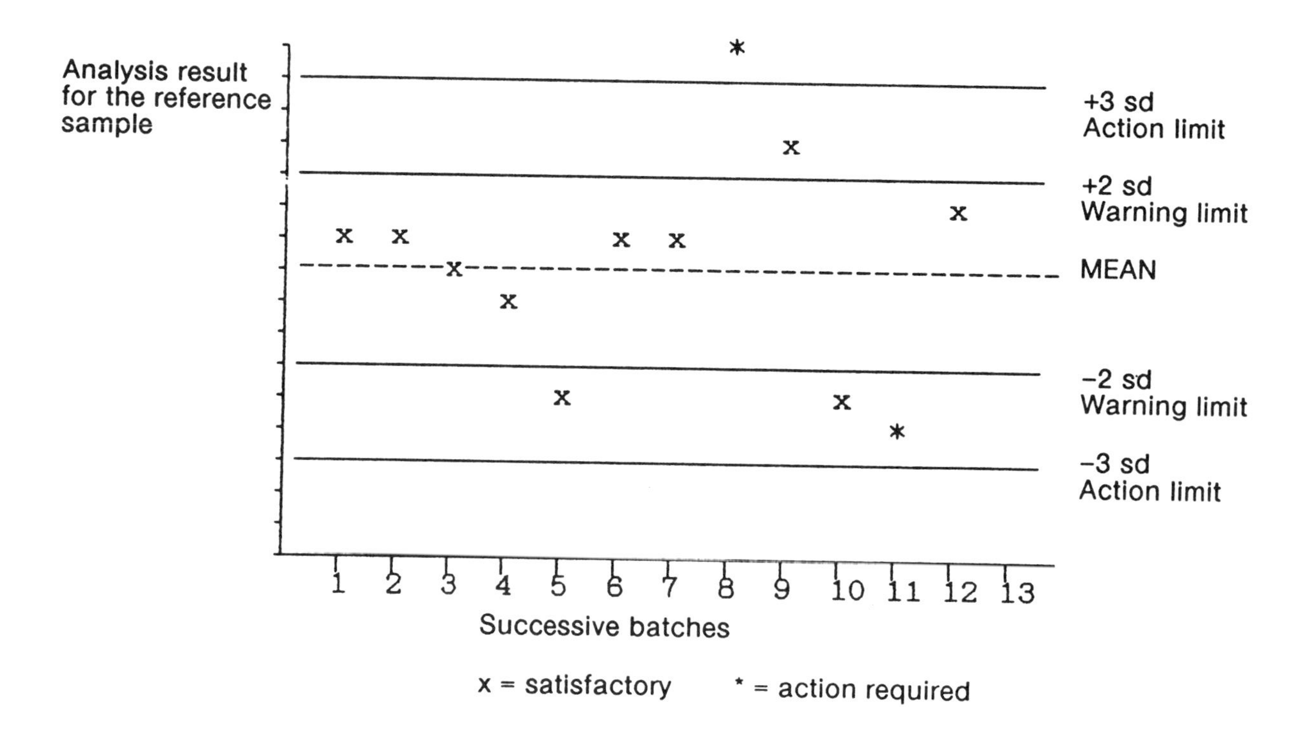

Figure 5.1 Quality Control Chart

5.10.4 Inter-laboratory standardisation

To compare data sets validly, all data should be obtained by using identical methods. With soil and plant analysis, the important aspect to standardise is the preparation of an analate (e.g. the extract or digest) rather than the quantitative determination. The former is where strict observance of the method protocol must be observed, whereas for the latter, any method that makes the quantitative determination accurately is satisfactory.

Standardisation between collaborating labs can be checked and improved by swapping reference materials, and then comparing results. Such materials are referred to as "external references".

The International Soil-Analytical Exchange programme offers an world-wide external reference service. For more information about this, contact ISE, c/o Dr V Houba, Dept of Soil Science and Plant Nutrition, Wageningen Agricultural University, PO Box 8005, 6700 EC Wageningen, Netherlands, tel +31 8370 82344, tlx 45015 intas nl, fax +31 8370 83766.

It is, however, strongly recommended that external references are also exchanged between "local" labs wherever possible.

Chapter 6 CHEMICAL ANALYSES

Note: Where "water" is specified in a procedure, it refers to either distilled or deionised water; where "tap" water is adequate, it is explicitly stated.

Note: Solution concentrations for colorimetric procedures are derived graphically in the following sections; alternative methods using regression are nevertheless equally appropriate.

6.1 WATER CONTENT DETERMINATION FOR DATA CORRECTION

Procedure

1 Weigh about 1 g of material to 0.001 g accuracy into a dry container of known weight (W1, also to 0.001 g accuracy). Record the total weight of the material plus container (W2).

Note: For some materials, e.g. field fresh soils or plant material, a much larger sample size should be taken. Weighings should be made to ± 0.1% accuracy.

2 Dry at 105°C for 2 hr, or until the weight stabilizes for larger sample sizes.

3 Allow to cool in a desiccator and reweigh the container plus dry material (W3).

Calculation

Water content (%) = (W2 - W3)/(W3 - W1) x 100

Dry mass (g) = {100/(100 + % water)} x (W2 - W1)

6.2 pH AND ELECTROCONDUCTIVITY SCREENING

Procedure

1 Add 50 ml water to 20 ± 0.1 g soil.

2 Stir the mixture for 10 min, leave to stand for 30 min, stir again for 2 min.

3 Measure the pH of the supernatant liquid.

For samples with pH < 6, analyse for exchangeable acidity, pH_{KCl} and ECEC. For samples with pH ≥6, analyse for EC and CEC.

4 Allow to settle for 1 hr then measure the electroconductivity (EC) of the supernatant liquid.

For samples with an EC > 1.0 mS/cm consider saturated paste extract conductivity analysis.

6.3 SATURATED PASTE EXTRACT CONDUCTIVITY

The electroconductivity screening will have identified any soils which are potentially saline. The electroconductivity of the saturated paste extract is measured to determine the level of salinity.

Reagents

Potassium chloride.

Standards

1 Dissolve 0.7456 g KCl in 1000 ml water: 1.412 mS/cm at 25°C.
2 Dissolve 7.456 g KCl in 1000 ml water: 12.900 mS/cm at 25°C.

Procedure

1 Weigh about 300 ± 25 g soil into a plastic container.

2 Add water to the soil with stirring until it is nearly saturated.

3 Allow the mixture to stand covered for several hours to permit the soil to imbibe the water, and then add more water to achieve a uniformly saturated soil-water paste. At this point the soil paste glistens as it reflects light, flows slightly when the container is tipped, slides freely and cleanly off a spatula, and consolidates easily by tapping or jarring the container after a trench is formed in the paste with the side of a spatula.

4 After mixing, allow the sample to stand (preferably overnight, but at least 4 hr), and then recheck the criteria for saturation. Free water should not collect on the soil surface, nor should the paste stiffen markedly or lose its glisten. If the paste is too wet, add additional dry soil to the paste mixture.

5 Transfer to a Buchner filter funnel fitted with Whatman No. 42 filter paper. Apply vacuum, and collect the filtrate. If the initial filtrate is turbid, refilter.

6 Measure the conductivity filtrate against that of the standards.

Reference

Rhoades, J.D. (1982) Soluble Salts. In: Page, A.L., Miller R.H. and Keeney D.R. (Eds.) *Methods of Soil Analysis, Part 2.* Second Edition. American Society of Agronomy, Inc., Madison.

6.4 EXCHANGEABLE CATIONS

The predominant exchangeable cations are either "basic" (K^+, Ca^{2+} and Mg^{2+}) or "acidic" (Al^{3+} and H^+). The "basic" cation exchange is not pH dependent and they are extracted with ammonium acetate at pH 7. The exchange of the "acidic" cations is pH dependent and they are extracted with an un-buffered potassium chloride solution.

Note: SI units are cmol(+)/kg. This is numerically equal to me/100 g.

6.4.1 Exchangeable bases

Reagents

Ammonium acetate, 1 M, pH 7.0: dissolve 77.1 g ammonium acetate in 950 ml water. Adjust the pH to 7.0 with acetic acid or aqueous ammonia. Make up to 1000 ml with water. Mix well.
Calcium carbonate
Hydrochloric acid, 2 M
Hydrochloric acid, conc.
Lanthanum chloride
Magnesium sulphate
Potassium chloride
Sulphuric acid, conc.

Standards

1 Stock standard solution (100 µg/ml K^+): dissolve 1.9067 g dry KCl in water and make up to 1000 ml. Dilute tenfold with water.
2 Stock standard solution (1000 µg/ml Ca^{2+}): dissolve 2.4973 g dry $CaCO_3$ in about 200 ml water containing 5 ml conc. HCl, boil to drive off CO_2, cool and dilute to 1000 ml in a volumetric flask with water.
3 Stock standard solution (100 µg/ml Mg^{2+}): dissolve 1.0136 g $MgSO_4.7H2O$ in water containing about 1 ml conc. H_2SO_4. Dilute to 1000 ml with water.
4 Lanthanum chloride solution (2000 µg/ml La^{3+}): dissolve 10.6939 $LaCl_3.7H_2O$ in water with aid of 1.6 ml 2 M HCl and dilute to 2000 ml. (Alternatively, take 40 ml 10% La solution and dilute to 2000 ml.)

For working standards:
5 Pipette 0, 1, 2 and 3 ml of each stock standard solution into 100 ml volumetric flasks.
6 Add 20 ml ammonium acetate solution and 20 ml lanthanum chloride to each flask and dilute to volume with water. Mix well.

Procedure

1 Plug the bottom of a 60 mm diameter funnel (capacity 25 ml) with cotton wool. Add 5 ± 0.1 g of soil. Leach with 10 successive 20 ml aliquots of 1 M ammonium acetate over a period of not less than 2 hours, collecting the leachate in a 250 ml volumetric flask.

Note: If the determination of CEC (for soils with pH ≥ 6) is required, retain the ammonium saturated soil in the leaching column for its CEC estimation.

2 Make the volumetric flask up to the 250 ml mark with ammonium acetate solution and mix well.

3 Pipette 20 ml of leachate into a 100 ml volumetric flask, add 20 ml lanthanum chloride solution and make up to the mark with water. Determine the K^+ content by flame emission spectroscopy, and Ca^{2+} and Mg^{2+} by atomic absorption spectroscopy.

Calculation

Determine graphically. For the 5.0 g soil:250 ml leachate ratio, diluted 5 times, the most concentrated working standards prepared above correspond to 1.2 K^+, 37.5 Ca^{2+} and 6.2 Mg^{2+} me/100 g soil.

6.4.2 Exchangeable acidity: Al^{3+} (+ H^+)

The exchangeable acidity in non-organic soils with a pH > 3 but < 5.5 will be almost completely comprised of exchangeable aluminium. For these soils it is therefore assumed that this extraction will also yield a value for exchangeable aluminium.

Reagents

Potassium chloride, 1 M: dissolve 74.55 g potassium chloride in 1000 ml water
Phenolphthalein solution: dissolve 1 g phenolphthalein in 100 ml ethanol
Sodium hydroxide, 0.1 M: dissolve 4.0 g sodium hydroxide in 1000 ml water

Procedure

1 Weigh 10 ± 0.1 g soil into a 100 ml beaker.

2 Add 25 ml 1 M potassium chloride.

3 Stir then leave for 30 min.

4 Filter through a Buchner funnel and leach with 5 successive 25 ml aliquots of 1 M potassium chloride.

5 Add 5 drops phenolphthalein solution and titrate with 0.1 M sodium hydroxide to the first permanent pink endpoint. Correct for a blank of sodium hydroxide titre on 150 ml potassium chloride solution.

Calculation

Exchangeable acidity (me/100g) = (ml NaOH sample - ml NaOH blank)

6.4.3 Cation exchange capacity (CEC)

Reagents

Ammonium acetate, 1M, pH 7.0: dissolve 77.1 g ammonium acetate in 950 ml water. Adjust the pH to 7.0 with acetic acid or aqueous ammonia. Make up to 1000 ml with water. Mix well.
Ethanol, 95%
Potassium chloride, 1 M: dissolve 74.55 g potassium chloride in 1000 ml water.

Procedure

1 Weigh 2.5 ± 0.01 g soil into a 50 ml centrifuge tube.

2 Add 33 ml 1 M potassium chloride, stopper the tube and shake for 5 min.

3 Un-stopper the tube and centrifuge until the supernatant is clear.

4 Discard the supernatant solution.

5 Repeat steps 2 - 4 four times.

6 Add 20 ml 95% ethanol, stopper the tube and shake for 5 min.

7 Un-stopper the tube and centrifuge until the supernatant is clear.

8 Discard the supernatant solution.

9 Repeat steps 6 - 8 two more times.

10 Add 33 ml 1 M ammonium acetate stopper the tube and shake for 5 min.

11 Un-stopper the tube and centrifuge until the supernatant is clear.

12 Pour off the supernatant solution into a 100 ml volumetric flask.

13 Repeat steps 10 - 12 twice, adding the clear supernatant to the volumetric flask.

14 Make the flask up to 100 ml with ammonium acetate and mix well.

15 Determine the potassium concentration in the volumetric flask as in Section 6.4.1, but do not dilute 5 times, i.e. omit step 3.

Note: If a centrifuge is not available, CEC can be determined by leaching; use t (see 6.4.1. above). Use the same quantities and reagents, but mix 5 g acid-washed sand with the 2.5 g soil to aid the ethanol elution steps.

Calculation

Determine graphically. For the 2.5 g soil:100 ml ratio and no dilution, the top standard will correspond to a CEC of 30 me/100g soil.

6.4.4 Effective cation exchange capacity (ECEC) and aluminium saturation

ECEC can only be used for soils which are not base saturated; i.e. pH < 7. ECEC gives the cation exchange capacity of the soil near its natural pH. (Methods such as ammonium acetate or barium chloride extractions determine the CEC of the soil at "artificial" pH values, and can seriously overestimate the ECEC of acid soils, particularly those dominated by variable charge clays). ECEC is calculated by summing the exchangeable cations determined from the two extractions given in Sections 6.4.1 and 6.4.2, i.e. exchangeable bases + exchangeable acidity. Aluminium saturation is the percentage of the ECEC occupied by exchangeable Al^{3+}.

ECEC (me/100g) = exch K^+ + exch Ca^{2+} + exch Mg^{2+} + exch acidity

Al^{3+} saturation (%) = (exch Al^{3+} / ECEC) x 100

6.5 SOIL ORGANIC MATTER AND ORGANIC CARBON

6.5.1 Definitions

Organic matter in soil is not a well-defined entity. It consists of a wide range of compounds forming a biochemical continuum from cellular fractions of higher plant, microbial and animal origin, through low and medium molecular weight organic substances of known structure, to high molecular weight humus compounds whose structure has yet to be characterised.

For the purposes of TSBF modelling and hypothesis testing the following fractions are defined:

Soil Organic Matter (SOM) is all the organic material in soil which has passed through a 2 mm sieve. It is measured by determining the organic carbon (OC) content of sieved soil. SOM is 40-60% carbon, depending on its composition and age. SOM is often assumed to be 58% OC (i.e. SOM = 1.724 x OC).

Microbial Biomass is the organic material in living bacteria, ascomycetes and fungi. It is determined by fumigation and comprises about 5% of the SOM. It is thought to be the most rapidly turning-over fraction of SOM, and is a major part of the "Active Pool" of the CENTURY model (Appendix L).

Soil Litter consists of organic particles smaller than 2 mm, but larger than 0.25 mm, which are only slightly transformed by decomposition. Consequently, their bulk density is relatively low (around 1.0 g/cm^3). The material which floats in water is the least altered. The material which does not float, but is caught on a 0.25 mm sieve, is thought to consist of fragments of resistant organic materials such as lignin, and may comprise most of the intermediate turnover rate pool of soil carbon (the "Slow Pool" in CENTURY). It is sometimes called "Light Fraction", since its bulk density is low in relation to the following pool. It contributes 45-65 % of the SOM.

Heavy Fraction consists of high molecular weight organic polymers, formed by microbial action on organic litters, in close physical association with clay and silt-sized soil particles. This association imparts a bulk density greater than 1.6 g/cm^3. It can be measured as the material which sinks in a 1.6 g/cm^3 fluid, but is more easily approximated as the OC content of dispersed soil passed through a 0.25 mm sieve. This is thought to represent the slowest turnover SOM component (the "Passive Pool" in CENTURY), and usually contributes 30-50 % of the SOM.

6.5.2 Total organic carbon in soils and soil extracts: background

The "wet" oxidation by acidified dichromate of organic carbon follows the reaction

$$2\ Cr_2O_7^{2-} + 3\ C^0 + 16\ H^+ = 4\ Cr^{3+} + 3\ CO_2 + 8\ H_2O$$

To ensure a complete oxidation of all organic C in the sample, it is necessary to heat the reaction at 150°C for 30 min; merely allowing the heat of dilution of the acid in the aqueous dichromate solution has been found to oxidize about 74% of organic C, averaged across many soil types.

Methods involving a partial oxidation of organic carbon, and hence a required factor in calculation, while less suitable, are acceptable where a heating block is not available; values for factors used should be quoted with the result. (In the calculation given in Section 6.5.3 a mean oxidation factor of 0.74 is used.)

The equation above shows that, given excess dichromate, the amount of organic C in the sample can be determined by measuring either the amount of unreacted dichromate (given a know initial amount), or by the amount of chromic (Cr^{3+}) produced. The former is determined by titration (i.e. as in Section 6.5.4), and the latter by colorimetry (i.e. as in Section 6.5.3). The titrimetric method is preferable for low concentrations of organic C, although the colorimetric method is very reproducible for samples with higher levels of organic C, and is very economical on technician time.

6.5.3 Total organic carbon in soils: colorimetric method

Suitable for all soils except with those where organic carbon <0.2%.

Reagents

Barium chloride, 0.4%: dissolve 4 g barium chloride in 1000 ml water.
Potassium dichromate, 5%: dissolve 50 g in 1000 ml water.
Sucrose
Sulphuric acid, concentrated (H_2SO_4, about 36 N).

Standards

1 Dry about 15 g sucrose at 105°C for 2 hr. Cool in a desiccator.
2 Dissolve 11.886 g dry sucrose in water and make up to 100 ml in a volumetric flask. This is a 50 mg/ml C solution.
3 Using a pipette transfer 0, 5, 10, 15, 20, 25 ml of the 50 mg/ml C stock solution into labelled 100 ml volumetric flasks and make up to the mark with water. Mix well. These are the working standards, and contain 0, 2.5, 5.0, 7.5, 10.0, 12.5 mg/ml C.
4 Pipette 2 ml of each working standard into labelled 100 ml conical flasks, and dry at 105°C, or into labelled digestion tubes. These now contain 0, 5, 10, 15, 20, 25 mg C.

Procedure without external heating

1 Weigh about 1 ± 0.001 g ground soil (<0.15 mm) into a labelled 100 ml conical flask. (If the soil is dark, or is suspected to be high in organic matter, use about 0.5 ± 0.001 g). Record the weight of soil, W.

2 Add 10 ml 5% potassium dichromate solution and allow it to completely wet the soil or dissolve the standards.

3 CAUTION: Add 20 ml H_2SO_4 from a fast burette and gently swirl the mixture.

4 Allow to cool, then add 50 ml 0.4% barium chloride, swirl to mix thoroughly, and allow to stand overnight, so as to leave a clear supernatant solution.

5 Transfer an aliquot of the supernatant solution into a colorimeter cuvette, and measure and record each standard and sample absorbance at 600 nm.

Procedure with external heating

1 Weigh about 1 ± 0.001 g ground soil (<0.15 mm) into a labelled 100 ml digestion tube. (If the soil is dark, or is suspected to be high in organic matter, use about 0.5 ± 0.001 g.) Record the weight of soil, W.

2 Add 2 ml water.

3 Add 10 ml 5% potassium dichromate solution and allow it to completely wet the soil or dissolve the standards.

4 CAUTION: Slowly add 5 ml H_2SO_4 from a slow burette and gently swirl the mixture.

5 Digest at 150°C for 30 min.

6 Allow to cool, then add 50 ml 0.4% barium chloride, swirl to mix thoroughly, and allow to stand overnight, so as to leave a clear supernatant solution.

7 Transfer an aliquot of the supernatant solution into a colorimeter cuvette, and measure and record each standard and sample absorbance at 600 nm.

Calculation

Plot a graph of absorbance against standard concentration. Determine solution concentrations for each unknown and the blanks. Subtract the mean blank value from the unknowns; this gives a value for corrected concentration, K. Where W = weight of soil:

For without external heating
% organic carbon = (K x 0.1) / (W x 0.74)

For with external heating
% organic carbon = (K x 0.1) / W

Reference

Baker, K.F. (1976) The determination of organic carbon in soil using a probe-colorimeter. *Laboratory Practice* **25**, 82-83.

6.5.4 Total organic carbon in soil extracts: titration method

Suitable for determinations of microbial biomass C, soil litter C and water soluble organic C, and for digests of soils very low in organic matter (>0.5%).

Reagents

Ferrous ammonium sulphate hexahydrate.
Potassium dichromate, 0.0667 M: dissolve 19.622 g dry potassium dichromate in 800 ml water, and dilute to 1000 ml.
Sulphuric acid, conc.
Acidified ferrous ammonium sulphate, 0.033 M: dissolve 12.940 g ferrous ammonium sulphate hexahydrate in 900 ml water, add 50 ml conc. sulphuric acid, allow to cool and make up to 1000 ml with water; mix well.
Indicator solution: dissolve 1.485 g *o*-phenantholine monohydrate (CAUTION: this is a poison) and 0.695 g ferrous ammonium sulphate hexahydrate in 100 ml of water.

Procedure

1 Transfer 4.00 ml of sample extract into a digestion tube.

2 Add 1.00 ml 0.0667 M potassium dichromate.

3 CAUTION: Add 5 ml concentrated sulphuric acid, mixing all the time.

4 Prepare 2 blank tubes (i.e. with reagents but without extracts).

5 Place the sample tubes and 1 blank in a preheated block at 150°C for 30 min, remove and allow to cool. Leave the other blank unheated.

6 Quantitatively transfer the tube contents to labelled 100 ml conical flasks, and add 0.3 ml (3 - 4 drops; not by mouth) indicator solution.

7 Using a magnetic stirrer to ensure good mixing, titrate all samples and blanks with acidified ferrous ammonium sulphate solution; the endpoint is a colour change from green/violet to red. Record the titres for each sample (ml_{sample}), the heated blank (ml_{HB}) and the unheated blank (ml_{UB}).

Note: The acidified ferrous ammonium sulphate solution needs to be standardised daily because of oxidation. This will establish its molarity, M:

1 Pipette 1 ml 0.0667 M potassium dichromate into a conical flask.

2 Add 0.3 ml (3 - 4 drops) indicator solution. (Do not use a mouth pipette.)

3 Titrate with acidified ferrous ammonium sulphate solution, and record the millilitres used (T).

Calculations

$M = 0.4 / T$

Organic carbon (%) = $\{(A \times M \times 0.003) / g\} \times (E / S) \times 100$

where
T = standardisation titre
M = molarity of ferrous ammonium sulphate ($\approx$ 0.033 M)
A = $(ml_{HB} - ml_{sample}) \times [(ml_{UB} - ml_{HB}) / ml_{UB}] + (ml_{HB} - ml_{sample})$
g = dry soil mass (g)
E = extraction volume (ml) (= 50 for above procedure)
S = digest sample volume (ml) (= 4 for above procedure)

Reference

Nelson D.W. and Sommers L.E. (1982) Total carbon, organic carbon and organic matter. In: Page, A.L., Miller, R.H. and Keeney, D.R. (eds.) *Methods of Soil Analysis,* Part 2. American Society of Agronomy, Madison. pp. 539-579.

6.5.5 Soil litter separation

Soil organic matter forms a continuum of particle size and density. All limits set for fractionation are arbitrary. For convenience it is recommended that techniques based on flotation in water are employed, rather than higher density fluids.

Higher carbon recoveries can be achieved by dispersion, followed by flotation in liquids of specific densities in the range 1.6 to 2.2 (Ford *et al.*, 1969; Turchenek and Oades, 1979), or by using density gradient centrifugation techniques or sonification (with or without sodium hexametaphosphate). Concentrations of organic C of soil litter separated in liquids of higher specific gravity may be more than 10 times those of the respective unfractionated soils, since they are more highly humified, but will contain increasing proportions of soil mineral components as well.

For TSBF studies the soil litter is defined as that material which passes a 2 mm, but not a 0.25 mm, sieve. This may be further subdivided into that which floats in water (less humified) and that which sinks (more humified). Separation of the soil litter can be achieved by using a commercially available root washing apparatus, for example as marketed by Mordue Bros., Newcastle, England or described by Smucker *et al.* (1982).

Procedure

1 Sieve 1 kg of soil through a 2 mm mesh sieve. Place the < 2 mm fraction in the root washer. It is a brass container fitted internally with baffles, and with water jets set horizontally at an angle of 5° with the container wall.

2 Allow mains pressure tap water to disperse the soil, so that soil litter and suspended fine organic-mineral components overflow through the opening at the apex of a brass cone centrally located within the apparatus. (In high clay content soils overnight soaking in 10% sodium hexametaphosphate may be required to aid dispersion prior to putting the soil in the brass container.)

3 Collect the soil litter on the removable sieve (0.25 mm) fitted to the bottom of the apparatus.

Note: In the absence of specialised apparatus the soil litter can be separated by agitating 250 g of soil (< 2 mm) in 5 - 10 litres tap water and carefully decanting the organic matter in suspension onto a 0.25 mm sieve. It is clearly difficult to standardise such a separation method and although the use of some design of fluidising column is preferable, this method gives a fairly good estimate; its comparative advantages are cheapness, simplicity and therefore wide applicability.

4 Wash the soil litter material collected on the 0.25 mm sieve with water (distilled) and filter using a Buchner funnel to remove excess water.

5 Dry the material at 85°C to constant weight.

6 Retain a subsample (of known mass) for possible N analysis if the soil litter is found to constitute > 20 % of the total organic C.

7 Combust the soil litter at 550°C for 6 hr in a muffle furnace. Results are expressed as ash-free organic matter.

Calculation

Ash-free mass (g) = dry mass - ash mass

If there is a large quantity of soil litter, only a subsample need be ashed. In this case,

Ash (%) = (ash mass/subsample mass) x 100 and

Ash-free mass (g) = dry mass x {(100 - ash %)/100}

Reference

Smucker, A.J.M., McBurney, S.L. and Srivastava, A.K. (1982) Quantitative separation of roots from compacted soil profiles by the hydropneumatic elutriation system. *Agronomy Journal* **74**, 500-504.

6.5.6 Microbial biomass

Biomass C denotes the amount of C in the total soil biota; it does not necessarily reflect their "activities". Its measurement does not require soil fractionation to obtain the soil biota as a separate entity (except perhaps the soil macrofauna), but rather the application of any of several indirect techniques (Jenkinson and Ladd, 1981) all of which give approximate measures only.

The chloroform fumigation-incubation technique (Jenkinson and Powlson, 1976; Anderson and Domsch, 1978) has been used extensively but there are problems using it for acid soils or soils which have recently had organic amendments. An alternative procedure in which the soil is extracted immediately after fumigation is recommended; it is quicker and simpler, and the results are highly correlated with those from the fumigation-incubation technique.

Reagents

Chloroform (alcohol-free): wash commercial chloroform with about 5% by volume conc. H_2SO_4 by shaking in a separating funnel. Separate off the acid, and wash the chloroform with 10 rinses of distilled water. (Store in the dark to prevent photochemical build-up of explosive by-products.)
Potassium sulphate, 0.5 M: dissolve 87.13 g potassium sulphate in 1000 ml water.

Procedure

Note: Up to 20 samples of field-fresh soils, taken a few days after soaking rain at the warmest time of the growing season, should be thoroughly mixed; subsamples of this composite field-fresh soil are required for the assay.

1 Sieve the soil to remove stones, coarse roots and all visible litter.

2 Weigh 2 subsamples of 10 ± 0.01 g of soil into 50 ml beakers and a third subsample of 10 ± 0.01 g soil into a 125 ml water-tight bottle.

3 Extract the sample in the bottle (t_0 sample) (see 6-8 below) while the first sample is being fumigated.

4 Determine % water content in one of the samples in the beaker (see Section 6.1) so as to be able to express the results on a dry-soil basis.

Fumigation

5 Place the beaker in a vacuum desiccator containing 30 ml alcohol-free chloroform in a shallow dish. Close the lid and apply the vacuum until the chloroform clearly evaporates. Close the tap on the desiccator and store in the dark for 5 days at 25°C. After 5 days, transfer the soil to a watertight 125 ml extraction bottle.

Extraction

6 Add 50 ml 0.5 M K_2SO_4 to the bottle, stopper tightly and shake for 30 min.

7 Filter the extract through a No. 42 Whatman filter paper, and retain the filtrate for analysis.

Analyse the extracts for dissolved organic C (as in Section 6.5.4; the external heating is not required as all the C is readily oxidisable):

Microbial biomass C = (Extracted C_{t1} - Extracted C_{t0}) x 2.64
(Vance *et al.*, 1988)

In addition, or alternatively, microbial N can be determined by analysing for total N in the extract after digestion (Section 6.6.1). Note that the N content in the digest is very low.

Microbial biomass N = (Extracted N_{t1} - Extracted N_{t0}) x 1.46
(Brookes *et al.*, 1985)

Microbial biomass P can be estimated using a procedure whereby inorganic P is extracted by 0.5 M sodium bicarbonate at pH 8.5. The extracted P is then determined by the ammonium molybdate-ascorbic acid method (see Section 6.10.5). It is desirable to correct for chloroform released P that is absorbed by soil during fumigation and extraction; an approximate allowance is made by incorporating a known quantity of P during extraction and then correcting for its recovery. After correcting for P extracted from unfumigated soil,

Microbial biomass P = (Extracted P_{t1} - Extracted P_{t0}) x 2.5

Note: It should be stressed that these methods are very sensitive to the analytical conditions, particularly with respect to the quality of the chloroform (it must be alcohol free for carbon determinations) and the temperature and duration of fumigation. The constants assumed in the calculations are approximate.

References

Amato, M. and Ladd, J.N. (1988) Assay for microbial biomass based on ninhydrin-reactive nitrogen in extracts of fumigated soils. *Soil Biology and Biochemistry* **20**, 107-114.

Anderson, J.P.E. and Domsch, K.H. (1978) Mineralisation of bacteria and fungi in chloroform-fumigated soils. *Soil Biology and Biochemistry* **10**, 215-221.
Brookes, P.C., Powlson, D.S. and Jenkinson, D.S. (1982) Measurement of microbial biomass phosphorus in soil. *Soil Biology and Biochemistry* **14**, 319-329.
Brookes, P.C., Landman, A., Pruden, G. and Jenkinson, D.S. (1985) Chloroform fumigation and the release of soil nitrogen: a rapid direct extraction method to measure microbial biomass nitrogen in soil. *Soil Biology and Biochemistry* **17**, 837-842.
Heanes, D.L. (1984) Determination of total organic-C in soils by an improved chromic acid digestion and spectophotometric procedure. *Communications in Soil Science and Plant Analysis* **15**, 1191-1213.
Hughes, S. and Reynolds, B. (1991) Effects of clearfelling on microbial biomass phosphorus in the Oh horizon of an afforested podzol in Mid-Wales. *Soil Use and Management* **4**, 183-188.
Jenkinson, D.S. and Powlson, D.S. (1976) The effects of biocidal treatments on metabolism in soil, V. A method for measuring soil biomass. *Soil Biology and Biochemistry* **8**, 209-213.
Jenkinson, D.S. and Ladd, J.N. (1981) Microbial biomass in soil: measurement and turnover. In: Paul, E.A. and Ladd, J.N. (eds.) *Soil Biochemistry* **5**, 415-471. Dekker, New York.
Vance, E.D., Brookes, P.C. and Jenkinson, D.S. (1987) An extraction method for measuring soil microbial biomass C. *Soil Biology and Biochemistry* **19**, 703-707.

6.6 NITROGEN

6.6.1 Digestion for total nitrogen (and phosphorus)

Analysis of total nutrients requires the complete breakdown or oxidation of organic matter. Wet oxidation is based on a Kjeldahl oxidation. Hydrogen peroxide is added as an additional oxidising agent, selenium takes the place of the traditional mercury catalyst and lithium sulphate is used to raise the boiling point. The main advantages of this method are that only the one digestion is required (for either soil or plant material) to bring nearly all of the nutrients into solution, no volatilisation of metals, nitrogen or phosphorus takes place and the method is simple and rapid.

Reagents

Hydrogen peroxide, 30%
Lithium sulphate
Selenium powder
Sulphuric acid, concentrated ($H_2SO_4 \approx 36N$)
Digestion mixture: add 0.42 g selenium powder and 14 g lithium sulphate to 350 ml 30% hydrogen peroxide and mix well. Slowly add with care 420 ml conc. H_2SO_4 while cooling in an ice bath. This mixture is stable for 4 weeks if stored at 2°C.

Note: Selenium is carcinogenic, so this reagent must be handled in a fume cupboard.

Procedure

1 Weigh about 0.2 ± 0.001 g ground soil (>0.15 mm) or plant material into a numbered digestion tube (min 75 ml size). Record the weight, W (g).

2 Add 4.4 ml digestion mixture to each tube. (**Note:** also digest 40 ml of blank digestion mixture for standard compensation; see Section 6.10.5)

3 Digest at 360°C for 2 hr. The solution should now be colourless and any remaining solids white. If colour can still be seen, heat for a further 1 hr.

Note: Even if the soil is low in organic carbon, and a shorter digestion time may therefore appear sufficient, boil for the full 2 hr to ensure that all the H_2O_2 is boiled off.)

4 Allow to cool.

5 CAUTION: Add about 50 ml water and mix well until no more sediment dissolves. Allow to cool.

6 Volumetrically make up to 100 ml with water and mix well.

7 Allow to settle so that a clear solution can be taken for analysis.

8 Determine the nitrogen in the digests as indicated in Section 6.6.2 (or 6.6.3) and phosphorus according to Section 6.10.5. Make up the working standards with the addition of 2.5 ml digested digestion blank.

6.6.2 Determination of nitrogen: distillation and titration

Free ammonia is liberated from solution by steam distillation in the presence of excess alkali. The distillate is collected in a receiver containing excess boric acid with an indicator (pH 4.5), and determined by titration.

Reagents

Ammonium sulphate
Boric acid
Hydrochloric acid
Sodium hydroxide
Sodium thiosulphate
pH 4.5 indicator solution
Standard ammonium sulphate (100 µg/ml NH_4^+-N): dissolve 0.4714 g dry ammonium sulphate in water and make up to 1000 ml in a volumetric flask.
Hydrochloric acid, M/140: prepare 0.1 M HCl and dilute to give M/140 (i.e. 0.00714 M). Standardise by distillation of 5.00 ml of the 100 µg/ml ammonium standard. For the titration, 1 ml of M/140 HCl will be equivalent to 0.1 mg NH_4^+-N.

Alkali mixture: dissolve 500 g NaOH and 25 g sodium thiosulphate in water with care. Cool and dilute to 1000 ml.

Boric acid indicator solution: dissolve 20 g boric acid in water, add 15 ml pH 4.5 indicator solution and dilute to 1000 ml.

Procedure

1 Set up a steam distillation apparatus. Use NH_3-free water if possible.

2 Pass steam through the apparatus for 30 min. Check the steam blank by collecting 50 ml distillate and titrating with M/140 HCl as given below. The steam blank should require no more than 0.2 ml acid.

3 Transfer an aliquot (with a volume of A ml) of sample solution to the reaction chamber and add 12 ml of alkali mixture. For digests use 25 ml of alkali mixture.

4 Commence distillation immediately and collect 25 ml distillate in a suitable receiver containing 5 ml of boric acid-indicator solution.

5 Titrate the distillate with M/140 hydrochloric acid to a grey end-point using a microburette. Record the volume of hydrochloric acid used (called the titre).

Occasionally check that the distillation recovery is satisfactory by taking an aliquot of the standard ammonium sulphate solution in place of the sample.

Calculation

Subtract the blank value from the sample titrations to give the corrected titre, T.

As 1 ml M/140 HCl = 0.1 mg NH_4^+-N then:

$$\text{Total N (\%)} = (T \times 0.1 \times 0.001^{*}) \times (S/A) / W \times 100/1$$
$$= (T \times S \times 0.01) / (A \times W)$$

where
* = conversion factor from mg to g
T = corrected titre (ml)
S = final digest solution volume (ml)
A = aliquot volume (ml)
W = sample weight (g)

Notes Deionised water is preferable to distilled water since the latter can give a high blank value. Ammonium ions can be removed from water by shaking with a strong cation exchange resin. Mains tap water may in fact have a lower NH_4^+-N content than distilled water in some areas.

For soil extracts or aqueous solutions take an aliquot of 50 ml.

The addition of the strong alkali liberates all inorganic N as NH_4^+. To determine NH_4^+-N only, use 0.2 g MgO instead of the alkali solution. The NO_3^+ can then be released in a second distillation after adding 0.4 g Devarda's alloy. In the case of acid digests, all the inorganic N will be in the NH_4^+ form, unless the NO_3^- has been protected (Bremner, 1982).

Reference

Bremner, J.M. (1982) Inorganic nitrogen. In: Page, A.L., Miller, R.H. and Keeney, D.R. (eds.) *Methods of soil analysis. Part 2.* Second Edition. American Society of Agronomy, Madison.

6.6.3 Colorimetric determination of ammonium

Reagents

Sodium citrate
Sodium hydroxide
Sodium hypochlorite solution, 5% available Cl^-
Sodium nitroprusside (CAUTION: poison)
Sodium salicylate
Sodium tartrate
Reagent N1: dissolve 34 g sodium salicylate, 25 g sodium citrate and 25 g sodium tartrate together in about 750 ml water. Add 0.12 g sodium nitroprusside and when dissolved make up to 1000 ml with water. Mix well.
Reagent N2: dissolve 30 g sodium hydroxide in about 750 ml water. Allow to cool, add 10 ml sodium hypochlorite solution and make up to 1000 ml with water. Mix well.

Note: Reagents N1 and N2 should be made at least 24 hours before use and stored in the dark.

Standards

1. Dry about 7 g ammonium sulphate at 105°C for 2 hr. Cool in a desiccator.
2. Dissolve 4.714 g dry ammonium sulphate in water and make up to 1000 ml in a volumetric flask. This is a 1000 μg/ml NH_4^+-N stock solution.
3. Pipette 50 ml of the 1000 μg/ml NH_4^+-N solution into a 500 ml volumetric flask and make up to the mark with water. This is a 100 μg/ml N solution.
4. Pipette 0, 5, 10, 15, 20 and 25 ml of the 100 μg/ml NH_4^+-N solution into labelled 100 ml volumetric flasks. The standards must be made up in exactly the same solution as the final samples, excluding the sample. For example, when determining the N content of digests, each standard must have 4.4 ml of digested digestion mixture (6.6.1) added before making up to the mark with water and mixing well.

These are the working standards and contain 0, 5, 10, 15, 20 and 25 μg/ml NH_4^+-N.

Procedure

1 Using a micropipette, transfer 0.100 ml of each standard and sample into suitably marked test tubes.

2 Add 5.00 ml of reagent N1 to each test tube, mix well and leave for 15 minutes.

3 Add 5.00 ml of reagent N2 to each test tube, mix well and leave for 1 hr for full colour development. The colour is stable for the day only.

4 Read each standard and sample absorbance at 655 nm.

Calculation

Plot a graph of absorbance against standard concentration. Determine solution concentrations for the sample and the blanks. Subtract the mean blank value from the sample; this gives a value for corrected concentration, C.

Nitrogen (%) = {(C x V) / W} x 0.0001

where

C = corrected concentration (μg/ml)
V = final digest or extract volume (ml)
W = weight of sample (g)

For the digestion procedure listed in Section 6.6.1:

Nitrogen (%) = (C / W) x 0.01

6.6.4 Colorimetric determination of nitrate

The determination of nitrate in soil usually follows an extraction in 0.5 M K_2SO_4. The procedure is to shake 10.0 g of fresh soil in 20 ml of extractant for 30 minutes at 60 rpm. Filter (use nitrate-free paper such as Whatman 42) or centrifuge the sample, and determine the nitrate in the clear solution. Do not extract in 1 M KCl, since the Cl^- ion can cause inference with the colorimetric reaction.

Freshly sampled soil must be used, since stored soil may have accumulated nitrate as a consequence of mineralisation. Correct the final value for the soil water content.

Reagents

Sodium hydroxide, 4 M: dissolve 160 g sodium hydroxide in 1000 ml water.
Salicylic acid, 5%: dissolve 5 g salicylic acid in 95 ml conc. sulphuric acid. (Make the day before; stable for 7 days if stored in a dark, cool place).

Standards

1 Dry about 10 g potassium nitrate at 105°C for 2 hr. Cool in a desiccator.
2 Dissolve 7.223 g dry potassium nitrate in water and make up to 1000 ml in a volumetric flask. This is a 1000 μg/ml NO_3^--N stock solution.
3 Pipette 25 ml of the 1000 μg/ml NO_3^--N solution into a 500 ml volumetric flask and make up to the mark with water. This is a 50 μg/ml NO_3^--N solution.
4 Pipette 0, 2, 4, 6, 8 and 10 ml of the 50 μg/ml NO_3^--N solution into labelled 50 ml volumetric flasks. Make up to the mark with extractant so that the standards and samples are in identical solutions, and mix well. These are the working standards and contain 0, 2, 4, 6, 8 and 10 μg/ml NO_3^--N.

Procedure

1 Micropipette 0.5 ml of each standard and sample into suitably marked test tubes.

2 Add 1.0 ml of salicylic acid solution to each test tube, mix well immediately the acid is added (use a vortex mixer, carefully) and leave for 30 minutes.

3 Add 10.0 ml of sodium hydroxide solution to each test tube, mix well and leave for 1 hr for full colour development. The colour is stable for 12 hr only.

4 Read each standard and sample absorbance at 410 nm.

Calculation

Plot a graph of absorbance against standard concentration.
Determine the solution concentrations for each unknown and the blanks. (If the extract is coloured, run a blank using extract, but use 1 ml sulphuric acid instead of the salicylic acid solution.) Subtract the mean blank value from the unknowns; this gives a value for corrected concentration, C.

NO_3^--N (μg/g soil)= (C x V) / W

where

C = corrected concentration (μg/ml)
V = extract volume (ml)
W = weight of sample (g)

Reference

Cataldo, D.A., Haroon, M., Schrader, L.E. and Youngs, V.L. (1975). Rapid colorimetric determination of nitrate in plant tissue by nitration of salicylic acid. *Communications in Soil Science and Plant Analysis* **6**, 71-80.

6.7 NITROGEN MINERALISATION

Note: The method for determining microbial biomass nitrogen is given in Section 6.5.6 (Microbial biomass).

6.7.1 Introduction

Net nitrogen mineralisation is often equated with nitrogen availability. No single method for estimating nitrogen availability has gained universal acceptance, and indeed it is unlikely that any single method will prove applicable to all sites and conditions (Bremner, 1965; Keeny, 1980).

In situ field methods employing the incubation of undisturbed cores theoretically offer the best estimate of nitrogen mineralisation. The method is however prone to problems of compaction and water-logging (both leading to denitrification), especially in clayey soils. The TSBF Programme recommends the field incubation method (Section 6.7.2) for process studies wherever possible. When field incubation is impossible, use the aerobic laboratory incubation as the next best alternative method (see Section 6.7.4). The anaerobic mineralisation index determined in the laboratory is the site characterisation method.

Laboratory methods involve an incubation under either anaerobic or anaerobic conditions. Aerobic methods are relatively simple, so many replicates can be performed. The method does however tend to over estimate N mineralisation rates, due to the massive disturbance (hence aeration) of the sample. It is however useful for studying the controls on N mineralisation between treatments, or when conditions preclude field estimations. The anaerobic method is preferable for site characterisation, and hence for comparisons between sites; it yields the anaerobic N mineralisation potential or "N mineralisation index". The anaerobic conditions prevent the oxidation of NH_4^+ to NO_3^-, resulting in a build up of the NH_4^+ which is then analysed.

6.7.2 Nitrogen mineralisation (field incubation method)

The modification of the *in situ* incubation method of Raison *et al.* (1987) outlined here, provides comparative estimates of N-mineralisation in sites with moderate to low annual rainfall. TSBF does not recommend using this method for estimating immobilisation, leaching and plant uptake as described in the original method of Raison *et al.* (1987).

The original method involved sampling and analysis of paired cores to accommodate the high spatial variability in N-mineralization. This proved very demanding in terms of sample handling and analysis. An alternative procedure is therefore recommended which involves random sampling in sub-lots and bulking each set of cores in each sub-plot to provide a single sample for analysis. This method provides a means of comparing sites and treatments.

Iron or plastic tubes (approximately 30 cm long and 50 mm internal diameter, or larger if compaction becomes a real problem) are used to take initial soil samples and to isolate soils during incubation. Galvanised piping should be weathered prior to use to avoid possible effects of zinc toxicity.

Procedure

1 Randomly insert 6 tubes into the soil in each sub-plot. The tubes should be driven or pushed in 25 cm if possible (to prevent root in-growth), leaving 5 cm of tube projecting above the soil surface.

2 Remove three of the tubes immediately, bulk the soils for each sub-plot (but see Note below), and determine initial mineral-N (NH_4^+ and NO_3^-) concentrations ($time_0$) (sections 6.6.3 and 6.6.4). Cover the remaining tubes with polyethylene (or a plastic cup) to protect the core from leaching effects by rain.

3 Sample the remaining three cores and determine mineral-N (NH_4^+ and NO_3^-) concentrations after a period of time which will have to be determined empirically for each site (1 to 2 weeks is a guideline).

Note: The cores can be stored at 4°C for a few days until processed, and can be sectioned by depth (0-10 and 10-25 cm) if a more detailed analysis is desired; this is essential if assaying treatments which affect the surface. The soils should be processed and analysed for mineral-N as described in Section 6.6.3 and 6.6.4. Net mineralisation is calculated as the difference in mineral-N between the two time points (time 1 - time 0) and is reported as μg N/g dry soil/time.

References

Bremner, J.M. (1965) Nitrogen availability indexes In: Black, C.A. (ed.), *Methods of Soil Analysis Part 2. Chemical and Microbiological Properties.* American Society of Agronomy Inc Monograph 10. Madison, Wisconsin, pp.1324-1345.

Keeny, D.R. (1980) Prediction of soil nitrogen availability in forest ecosystems: a literature review. *Forest Science* **26**, 159-171.

Matson, P.A. and Vitousek, P.M. (1981) Nitrogen mineralisation and nitrification potentials following clearcutting in the Hoosier National Forest, Indiana. *Forest Science* **27**, 781-791.

Raison, R.J., Connell, M.J. and Khanna, P.K. (1987) Methodology for studying fluxes of soil mineral-N *in situ*. *Soil Biology and Biochemistry* **19**, 521-530.

Smethurst, P.J. and Nambiar, E.K.S. (1989) An appraisal of the *in situ* core technique for measuring nitrogen uptake by a young *Pinus radiata* plantation. *Soil Biology and Biochemistry* **21**, 939-942.

Waring, S.A. and Bremner, J.M. (1964) Ammonium production in soil under waterlogged conditions as an index of nitrogen availability. *Nature* **201**, 951-952.

6.7.3 Aerobic incubation

Two series of aerobic incubations are conducted, one at ambient field moisture, the other at the best approximation of field capacity, i.e. non-limiting water availability. (In practice disturbed soils will have a lower water holding capacity than undisturbed cores.) The first series provides an estimate of net nitrogen mineralisation at the time of sampling and reflects the effect of moisture status, both from seasonality and treatment related effects. The second series provides an estimate of mineralisation under optimum conditions and reflects the effect of substrate quality on mineralisation (Matson and Vitousek, 1981).

Soils used for the incubations should be composite, or bulked, samples obtained from 10-20 soil cores randomly sampled within a specific area or transect. A minimum of 10 composite samples per hectare is recommended. Sampling depth should be consistent with that of other soil measurements; however the surface soil (0-10, 0-15 or 0-20 cm) should be incubated separately from lower depths to enable detection of differences between treatments.

Reagents:

Potassium sulphate, 1 M: dissolve 174.25 g potassium sulphate in 1000 ml water.

Procedure:

1. Bulk soil cores and take subsamples for immediate gravimetric determination of moisture content (see Section 6.1). The remaining material is sorted to remove stones and as much root material as possible without disrupting gross aggregate structure and allowing the soil to dry.

2. For each composite soil sample weigh 50 g into each of two flasks or plastic bags.

3. Add water to one sample to adjust the gravimetric water content to the best approximation of field capacity (as determined in Section 7.3), and drain excess water; no water is added to a second sample.

4. Plug the flasks (or bags) with cotton wool (or if plastic bags are used they are tied off), weigh and incubate at 26 ± 2°C for 14 days. The weight of the flasks should be checked periodically and corrections made for water loss. (If the temperature is not within the specified range, the actual temperature mean and range should be reported.)

5. At the beginning of the incubation ($time_0$) extract a further subsample of 5 g in 1 M potassium sulphate for 30 minutes. A soil to solution ratio of 1:10 is often recommended but it may be necessary to use less extractant if nitrate and ammonium levels are low.

6. Filter the extract. **Note:** all filter papers are contaminated with some ammonium so it is best to pre-rinse filters by passing through 50 ml of distilled water.

7. Determine mineral NH_4^+-N and NO_3^--N in the filtered extract as described in Sections 6.6.3 and 6.6.4.

8. At the end of the incubation period ($time_1$) the two incubated soils are extracted and analysed in the same manner as for $time_0$.

Net mineralisation is calculated as the difference in mineral-N between the two time periods (time 1 - time 0). Results are expressed as μg N/g dry weight soil/14 days.

6.7.4 Anaerobic N mineralisation index

This determination should be conducted on "field-fresh" soils, preferably within 1 hr of sampling; about 100 g of stone-, root- and large litter-free sample should be returned to the laboratory. (If it is not possible to analyse with 1 hr, transport the soils in a cold box.) The standard 40°C/7 day incubation is followed:

Reagents

Potassium chloride, 1 M: dissolve 74.55 g KCl in 1000 ml water.
Potassium chloride, 2 M: dissolve 149.1 g KCl in 1000 ml water.

Procedure

1. Weigh one 100 ml airtight bottle (bottle A: W1) and weigh one 15 ml airtight bottle/polytop (bottle B: W3).

2. Add a 10 ml scoop of sample to bottle A and weigh (W2) and a 5 ml scoop of sample to bottle B and weigh (W4).

3. Determine the water correction factor on a third subsample as described in Section 6.1.

4. Add 50 ml 1 M KCl to bottle A and shake for 20 min at about 60 Hz.

5. Meanwhile add 12.5 ml water to bottle B, swirl to remove air bubbles, stopper, and place in an incubator (oven) set at 40°C.

6. After the 20 min extraction, determine $time_0$ NH_4^+-N in bottle A as described in Sections 6.6.2 or 6.6.3. (An ammonia specific electrode may be used: add 2 ml 5 M NaOH to the sample after the shaking, immediately insert the electrode and read exactly 2 min after electrode exposure.)

7. After 7 days, remove bottle B from the incubator.

8. Quantitatively transfer with a known volume of 2 M KCl to a clean container and shake for 20 min at about 60 Hz.

9. After the 20 min extraction, determine $time_1$ NH_4^+-N in bottle B as described in Sections 6.6.2 or 6.6.3. (An ammonia-specific electrode may again be used as in step 6 above).

Calculation

soil mass $time_0$ (g) = W3 - W1
soil mass $time_1$ (g) = W4 - W2 (Express the results on a dry soil basis.)

Anaerobic N mineralisation rate (μgN/g soil/day) = $\{(time_1\ NH_4^+\text{-N}) - (time_0\ NH_4^+\text{-N})\}\ /7$

6.8 DENITRIFICATION

6.8.1 Introduction

The term denitrification mainly covers the reduction of nitrate and nitrite to gaseous products such as nitrous oxide and molecular nitrogen. The process is carried out by a diverse group of bacteria using nitrogen oxides as terminal electron acceptors in lieu of oxygen under anaerobic conditions. The reductive pathway is generally accepted to be:

NO_3^- --------> NO_2^- --------> NO --------> N_2O --------> N_2

Other pathways include the direct oxidation of ammonium to nitrous oxide during autotrophic nitrification (Yoshida and Alexander, 1970) and by nitrate-respiring bacteria which reduce nitrite to ammonium (Focht and Verstraete, 1977; Ingraham, 1981). The nitrate-respiring bacteria are of interest because the disappearance of nitrate might otherwise be ascribed to denitrification and because they effectively reduce the loss of nitrate which accumulates in excess of plant demand.

A discussion of the methods for the quantification of nitrification can be found in Appendix F ("Nitrogen availability") or in Payne (1991). The acetylene inhibition method represents the most direct way of measuring denitrification in field and laboratory studies. Laboratory studies have shown that 0.1 - 10% (v/v) acetylene effectively blocks N_2O, and is easily measured against its background concentration in the atmosphere. This avoids the restrictions of having to use mass spectrometry and ^{15}N tracer techniques to determine dinitrogen emissions from soils.

6.8.2 Procedure

Reagents

Calcium carbide

Procedure

1 Take soil cores in either a plastic tube housed inside a steel tube that is driven into the soil or in a steel tube with the core carefully transferred to a plastic tube of larger diameter immediately after sampling. Care must be taken to maintain the existing core structure as completely as possible.

2 Within a few hours of collection, stopper each end of the plastic tube with a septum and flush the tube with an equal volume of air from a large syringe to dispel accumulated N_2O.

3 Add sufficient acetylene (generated from calcium carbide plus water in a stoppered, evacuated flask, as opposed to coming from a commercial cylinder) to make up a 10% v/v (10 kPa) atmosphere. To ensure thorough acetylene distribution through the soil,

pump the tube atmosphere with a large syringe to alternately reduce and increase pressure in the soil pore space. This should also be done prior to each gas sampling.

4 Incubate the cores either in the laboratory at field temperature or in the field. Take gas samples from every core at least twice over a 12 hr incubation period, e.g. at 2 hr and 12 hr following the addition of acetylene. A more conservative procedure is to take gas samples at 2 hr intervals over this period. Incubation periods longer than 12 hr face potential problems with decreased nitrate and O_2 concentrations; decreased nitrate levels will tend to depress denitrification rates later in the incubation period, while decreases in O_2 concentrations will tend to accelerate rates.

5 Store samples in glass syringes or preferably in 3 ml pre-evacuated Venoject (TM) vials (Terumo Scientific, New Jersey, USA) until analysis; to avoid sample contamination it is best to overpressure sample vials by adding 5 ml of sample. If stored for longer than a few hours, vials should be checked for leakage (lack of positive pressure) before analysis.

6 Analyse N_2O (and if desired CO_2) by gas chromatography using a ^{63}Ni electron capture detector.

7 After the final sampling estimate the volume of the head+pore space in each tube and oven-dry an aliquot of the soil to determine water content. Estimate head+pore volume from the bulk density of the soil. (A more accurate head+pore space volume can be estimated by injecting a known quantity of air into the sealed tube and measuring the pressure difference with a pressure transducer; differences in pressure among tubes will be proportional to the pore+air space present. A tube filled with different quantities of water can serve to calibrate the transducer to different head+pore space volumes.)

Calculation (example)

The rate of denitrification is usually calculated on an area basis applying corrections for the solubility of N_2O in soil water using the coefficients in Table 6.1. This correction is based on the assumption that most N_2O production is initiated during incubation with acetylene and the water phase is then in equilibrium with higher N_2O concentrations than when collected from the field.

For example, soil cores containing 20 ml water incubated at 25°C with a head+pore space of 200 ml have a water/atmosphere quotient of 20/200 (= 0.1), and hence a correction factor of 0.059. If the rate of N_2O production was 500 ng/tube/12 hr then total N_2O production would be 500 + (500 x 0.059) = 44.1 ng/tube/hr.

Express the final result in ng/g oven-dry soil/hr.

Table 6.1 Correction factors for the solubility of N_2O in soil water in closed incubation flasks (Moraghan and Buresh, 1977).

Temperature	Water volume / Atmosphere volume					
°C	0.1	0.25	0.5	0.75	1.0	1.5
10	0.091	0.228	0.455	0.683	0.910	1.37
15	0.078	0.194	0.389	0.584	0.778	1.17
20	0.068	0.169	0.338	0.507	0.676	1.01
25	0.059	0.149	0.297	0.446	0.594	0.89
30	0.053	0.133	0.265	0.398	0.530	0.80

References

Focht, D.D and Verstraete, W. (1977) Biochemical ecology of nitrification and denitrification. In: Alexander, M. (ed.), *Advances in Microbial Ecology* **1**, 135-214. Plenum Press, New York.

Ingraham, J.L. (1981) Microbiology and genetics of denitrifiers. In: Delwiche, C.C. (ed.), *Denitrification, Nitrification and Atmospheric Nitrous Oxide*. John Wiley, New York, pp.45-65.

Moraghan, J.T. and Buresh, R. (1977) Correction for dissolved nitrous oxide in nitrogen studies. *Soil Science Society of America Proceedings* **41**, 1201-1202.

Payne, W.J. (1991) A review of methods for field measurements of denitrification. *Forest Ecology and Management* **44**, 5-14.

Yoshida, T. and Alexander, M. (1970) Nitrous oxide formation by *Nitrosomonas europaea* and heterotrophic microorganisms. *Soil Science Society of America Proceedings* **34**, 880-882.

6.9 PHOSPHORUS

Note: The method for determining microbial biomass phosphorus is given in Section 6.5.6.

6.9.1 Total phosphorus

The digestion procedure for soil and plant material is described in Section 6.6.1. For plant material, 100% recovery of P is achieved, as all the material is solubilised; for soils, the % recovery varies, but is very high if the solid material remaining after digestion is white.

Determine P in the digests as in Section 6.9.5.

Calculation

P in sample (%) = (C / W) x 0.01

where
W = weight of sample.

6.9.2 Bicarbonate extractable phosphate

Many extraction techniques for "plant-available" phosphate have been developed; TSBF recommends the bicarbonate extraction because of its suitability over a wide range of soil types and pH values.

Reagents

Sodium bicarbonate, 0.5 M, pH 8.5: dissolve 84 g sodium bicarbonate in about 1000 ml water. Make up to nearly 2000 ml with water, adjust the pH to 8.5 with 10% NaOH, mix and make up to 2000 ml.

Procedure

1 Weigh 2.5 ± 0.01 g soil (W) into a polyethylene bottle.

2 Add 50.0 ml extracting solution.

3 Shake the bottle for 30 min then filter through Whatman No. 42 filter paper.

4 Determine the bicarbonate extracted phosphate in the filtrate as in Section 6.9.5, but make the working standards in bicarbonate extracting solution; i.e. use extracting solution instead of water in steps 3 and 4.

Calculation

Bicarbonate extractable phosphate (μg/g) = (C x 20)/W

Reference

Watanabe, F.S. and Olsen, S.R. (1965) Test of an ascorbic acid method for determining phosphorus in water and $NaHCO_3$ extracts from soil. *Soil Science Society of America Proceedings* **29**, 677-678.

6.9.3 Resin extractable phosphate

A suitable measure of labile P is resin extractable phosphate. This is closely related to isotopically exchangeable P and therefore with the P pool that is in equilibrium with the soil solution.

Procedure for making resin bags

1 Prepare small resin bags (about 4 cm x 3 cm) by folding polyester screening (e.g. Estal Mono PE 400μm) and sealing the edges of the bag in a cool flame. If you are sewing the bags make sure you use polyester thread; other threads may rot.

2 Sieve a quantity of any gel-type strongly-basic Type I anion exchange resin through a 350 μm sieve.

3 Place sufficient of the >350 μm fraction in each bag to provide a total anion exchange capacity of about 10 mmol. This is about 3.5 g if Dowex 1-X 8 resin is used.

4 Heat seal or sew the open end of the bags closed, perpendicular to the closure at the opposite end, so as to form a tetrahedral shape which does not compress the resin (see Figure 6.1).

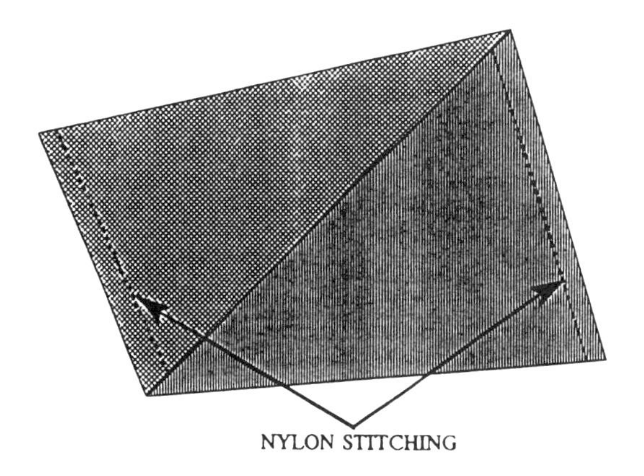

Figure 6.1 Tetrahedron resin bag; finished shape.

The resin in the bags is regenerated after each use and can be used many times. Resin bags must always be stored in slightly acidified water between use. (Resin membranes are now available (BDH, UK) which only need to be cut into strips and are much easier to use, but their extractive power is slightly different and they would need to be calibrated.)

Resin bag preparation

The best P extraction is obtained if the resin is in a mixed chloride-bicarbonate form:

Reagents

Sodium bicarbonate, 0.5 M: dissolve 42.0 g $NaHCO_3$ in 1000 ml water.

Procedure

1 Place the bags in a large beaker containing 100 ml 0.5 M $NaHCO_3$ per bag for 30 min, stirring occasionally. Repeat with fresh 0.5 M $NaHCO_3$. Wash the bags twice for 30 min each time in distilled water, and store them in the final wash water until use. They should be used within 24 hr.

Extraction

Note: When trying to analyse small quantities of P it is extremely important that the glassware is clean. Acid-wash glassware in dilute HCl solution if you suspect contamination; avoid using plastic containers if possible.

Reagents

Hydrochloric acid, 0.5 M.

Procedure

1 Weigh up to 4.0 ± 0.01 g (depending on P content) of sieved soil (< 2 mm) into a 50 ml bottle with a tightly fitting lid.

2 Add 40 ml of water and one resin bag, put on the lid and gently shake for about 14 hr, i.e. overnight.

3 Remove the resin bag, wash free of soil with distilled water and shake for 30 min with 20 ml of 0.5 M HCl. (Allow the bubbles to escape before putting the lids back on and shaking.)

4 Determine the phosphate content of the acid by the colorimetric procedure in Section 6.9.5 but make the working standards in 0.5 M HCl; i.e. use 0.5 M HCl instead of water in steps 3 and 4.

Wash the bags a few times in distilled water to remove the acid and then regenerate the resin (when required) as described above.

Calculation

Resin extractable phosphate (μg/g) = (C x 20) / W

where
W = weight of sample.

Reference

Sibbesen, E. (1978) An investigation of the anion exchange resin method for soil phosphate extraction. *Plant and Soil* **50**, 305-321.

6.9.4 Organic phosphorus

Soil organic matter is destroyed by igniting the soil sample at 550°C in a muffle furnace. This renders the organic-P acid extractable and the difference in the acid extractable P of an ignited and an unignited sample of the same soil is a measure of the total organic-P in the soil.

Reagents

Sulphuric acid, 1 M

Procedure

1 Weigh 1 ± 0.001 g of ground soil into a porcelain crucible, and place the crucible in a cool muffle furnace.

2 Slowly raise the temperature of the furnace to 550°C over a period of 1 to 2 hr.

3 Ash at 550°C for 1 h, allow the crucible to cool, and transfer the ignited soil to a 100 ml polypropylene centrifuge tube.

4 Weigh 1 ± 0.001 g of ground unignited soil into a separate 100 ml polypropylene centrifuge tube.

5 Add 50 ml of 1 M H_2SO_4 to both samples, and shake the tubes overnight.

6 Centrifuge or filter (Whatman No. 42, washed with 1M H_2SO_4) the samples to clear the solution.

7 Determine P in the extracts as in Section 6.9.5, but make the working standards in 1 M H_2SO_4; i.e. use 1 M H_2SO_4 instead of water in steps 3 and 4.

Calculation

$P_{ignited}$ (%) = (C x 0.005) / W
$P_{unignited}$ (%) = (C x 0.005) / W

Organic phosphorus (%) = $P_{ignited} - P_{unignited}$

where

W = weight of sample.

References

Bowman, R.A. (1989) A sequential extraction procedure with concentrated sulphuric acid and dilute base for soil organic phosphorus. *Soil Science Society of America Journal*, **53** 362-366.

Olsen, S.R. and Sommers, L.E. (1982). Phosphorus. In: Page, A.L., Miller, R.H. and Keeney, D.R. (eds.). *Methods of Soil Analysis,* Part 2. Second Edition. American Society of Agronomy, Inc., Madison.

6.9.5 Colorimetric determination of phosphorus

Reagents

Ammonium molybdate
Antimony sodium tartrate
Sulphuric acid, conc.
Ascorbic acid, 1%: dissolve 1 g ascorbic acid in 100 ml water; make a fresh solution daily.
Molybdate reagent: dissolve 4.3 g ammonium molybdate in 400 ml water in a 1000 ml measuring cylinder. Dissolve 0.4 g antimony sodium tartrate in 400 ml water, then add to the measuring cylinder. CAUTION: add carefully, with stirring, 54 ml H_2SO_4. Allow to cool and make up to 1000 ml with water. Mix well. Stable for 4 weeks at 2°C.

Standards

1 Dry about 7 g KH_2PO_4 at 105°C for 2 hr. Cool in a desiccator.
2 Dissolve 4.394 g dry KH_2PO_4 in water and make up to 1000 ml in a volumetric flask. This is a 1000 µg/ml P stock solution.
3 Pipette 10 ml of the 1000 µg/ml P solution into a 500 ml volumetric flask and make up to the mark with water (or with the appropriate solution as required for a given determination). This is a 20 µg/ml P solution.
4 Pipette 0, 5, 10, 15, 20 and 25 ml of the 20 µg/ml P solution into labelled 100 ml volumetric flasks. Make up to the mark with water (or with the appropriate solution as required for a given determination) and mix well. These are the working standards and contain 0, 1, 2, 3, 4 and 5 µg/ml P in suitable background solution.

Procedure

1 Pipette 1 ml standard or sample into a test tube.

2 Add 4.0 ml ascorbic acid solution.

3 Add 3.0 ml molybdate reagent and mix well.

4 Leave for 1 hr for the colour to develop fully.

5 Read the standard and sample absorbances at 880 nm.

Calculations

Plot a graph of absorbance against standard concentration.
Determine solution concentrations for each unknown and the blanks.

$$C\ (P\ \mu g/ml) = P_{sample} - P_{mean\ blank}$$

Calculations for the various P extracts are given in the appropriate sections above, where the derived value for C is used.

6.10 POLYPHENOLICS

The following method is adapted from King and Heath (1967) and Allen *et al.* (1974) to analyse for total extractable polyphenolics. Total soluble polyphenolics are analysed by the Folin-Denis method, and include hydrolysable tannins and condensed tannins, as well as non-tannin polyphenolics.

Swain (1979) reports extraction from 30 to 95% of total polyphenolics; this method therefore gives a very rough estimate of polyphenolics and it includes polyphenolics with varying reactivities and functions. Analysis of the extract using a protein-binding procedure such as the Bovine Serum Albumin method (Dewra *et al.*, 1988) can give an estimate of the reactive phenolics.

Reagents

Methanol, 50%
Orthophosphoric acid
Phosphomolybdic acid
Sodium carbonate, 17%: dissolve 17 g sodium carbonate in 100 ml water.
Sodium tungstate
Tannic acid
Folin-Denis reagent: add 50 g sodium tungstate, 10 g phosphomolybdic acid and 25 ml orthophosphoric acid to 375 ml water. Reflux for 2 hr (with glass beads in the flask to prevent superheating). Cool and dilute to 500 ml with water.

Standards

1 Tannic acid, 0.1 mg/ml: dissolve 0.050 g tannic acid in 500 ml of water in a volumetric flask, and make up to volume.
2 Using 0, 1, 2 and 4 ml of the tannic acid standard instead of the 1 ml of unknown, follow the procedure below from step 4.

Procedure

1 Weigh about 0.2 ± 0.001 g (W) of material into a 50 ml beaker.

2 Add 20 ml 50% methanol, cover with parafilm and place in a water bath at 77-80°C for 1 hr. (Note: Initially add about 1 ml of the methanol and stir with a glass rod to wet and disperse the finely ground material, to prevent lumping.)

3 Quantitatively filter (Whatman No 1) the extract into a 50 ml volumetric flask using 50% aqueous methanol to rinse, and make up with water. Mix well.

4 Pipette 1 ml of the unknown or standard into a 50 ml volumetric flask, add 20 ml water, 2.5 ml Folin-Denis reagent and 10 ml sodium 17% carbonate.

5 Make to the mark with water, mix well and stand for 20 min.

6 Read the standard and unknown absorbances at 760 nm.

Calculation

Plot a graph of absorbance against standard concentration.

Determine solution concentrations for each unknown and the blanks. Subtract the mean blank value from the unknowns; this gives a value for corrected concentration, C.

Total extractable polyphenolics (%) = (C x 5) / W

References

Allen, S.E., Max Grimshaw, A.H., Parkinson, J.A. and Quarmy, C. (1974) *Chemical Analysis of Ecological Materials.* Wiley, New York.

Dawra, R.K., Makkar, H.S.P. and Singh, B. (1988) Protein-binding capacity of microquantities of tannins. *Analytical Biochemistry* **170,** 50-53.

King, H.G.C. and Heath, G.W. (1967) The chemical analysis of small samples of leaf material and the relationship between the disappearance and composition of leaves. *Pedobiologia* **7**, 192-197.

Swain, T. (1979) Tannins and Lignins. In: Rosenthal, G.A. and Janzen, D.H. (eds.), *Herbivores: Their Interactions with Secondary Plant Metabolites.* Academic Press, New York.

6.11 LIGNIN AND CELLULOSE

6.11.1 Introduction

Lignin has been selected as the variable to assess the quality of the organic matter in the litter. Lignin is defined as the residual organic fraction after chemical extraction which is resistant to microbial degradation. Chemical methods for fractionating lignins are in general time consuming, labour intensive and subject to interferences. Each type of method is briefly considered and recommendations for procedures and sample handling are given.

Klasson methods (e.g. Browning, 1967) have been widely adopted and are based on the removal of the cellulose fraction by hydrolysis with 72% sulphuric acid. Removal of fats/lipids, however, using alcohol and benzene or ether, soluble carbohydrates with water, starch with dilute acid and protein with enzymes may be required prior to hydrolysis with ash and protein correction following analysis. This method, which has usually been regarded as the standard procedure, is unsuitable for large numbers of samples, but the reference has been included (Allen, 1974), should it be required for comparison and calibration purposes. Other schemes (e.g. Southgate, 1967) which remove and quantify each organic fraction in turn are also unsuitable for large scale use. An extraction method using acetyl bromide and UV measurement (Morrison, 1972) is rapid and has been used for forage crops. This has not found wide applications, is subject to interferences and therefore cannot be recommended for this study.

The chosen TSBF method is based on the acid detergent fibre method (ADF) (Van Soest, 1963) which has been widely adopted by agricultural analysts to provide a crude measure of resistant organic material. The residue may be further separated into lignin, cellulose and ash. Alternative lignin isolation procedures proposed by Van Soest, namely permanganate oxidation or sulphuric acid, were evaluated for a range of material (Rowland and Roberts, 1994). The method based on the destruction of cellulose with 72% H_2SO_4 proved robust and applicable to decomposition studies. This method has been used very successfully in laboratories operating within the TSBF programme. As the method does however involve multiple steps, with concomitant problems of reproducibility, formal TSBF collaborators are encouraged to use a common external laboratory for lignin analysis.

For non-formal TSBF collaborators, and if resources are available, the full method for lignin and cellulose analysis on ADF should be followed. Alternatively, the ADF fraction could be determined for all samples, and a limited number of bulked samples examined for lignin and cellulose.

6.11.2 Lignin and cellulose via acid detergent fibre (ADF)

Acid detergent fibre is prepared from plant material by boiling with a sulphuric acid solution of cetyltrimethyl ammonium bromide (CTAB) under controlled conditions. The CTAB dissolves nearly all the nitrogenous constituents and the acid hydrolyses the starch to leave a residue containing lignin, cellulose and ash (Clancy and Wilson, 1966). Cellulose is destroyed by 72% H_2SO_4; lignin is then determined by weight-loss upon ashing.

Reagents for ADF

Acetone (CAUTION: highly flammable)
Sulphuric acid/CTAB solution: dissolve 100 g cetyltrimethyl ammonium bromide in 5 litres 0.5 M sulphuric acid. Filter if cloudy.

Procedure for ADF

1 Weigh about 0.5 ± 0.001 g air-dry plant material into a 250 ml wide-neck conical flask. (Note: Use over-dry material if insufficient sample is available for % water content determination - see Section 6.1); air-dry is recommended as oven-drying *may* alter the component structure.)

2 Add 100 ml sulphuric acid/CTAB solution.

3 Gently reflux for 1 hr on a hot plate (top up with CTAB if volume deceases).

4 Filter hot through a pre-ignited sinter (No. 2) of known weight (W2), under gentle suction.

5 Wash the residue with 3 x 50 ml aliquots of boiling water.

6 Wash with acetone until no more colour is removed, and suck the fibre dry.

7 Dry in an oven at 105°C for 2 hr, cool in a desiccator and re-weigh (W3).

Calculation

Ash-containing ADF (%) = {(W3 - W2) / W1} x 100 x {100 / 100 - % water content}

Reagents for lignin and cellulose

Sulphuric acid (72% w/v): CAUTION: Add 720 ml concentrated H_2SO_4 to 540 ml water (add acid in small portions to water whilst cooling the vessel.)
Acetone (CAUTION: highly flammable)

1 Half-fill the sinter containing the ADF with cooled (15°C) 72% H_2SO_4.

2 Stir with a glass rod to produce a smooth paste and place the sinter in a suitable vessel to catch the waste acid as it drains away.

3 Refill the sinter with H_2SO_4 at hourly intervals.

4 After 3 hr, filter off the acid under vacuum and then wash the residue with hot water until the filtrate is free of acid. Wash off the glass rod.

5 Rinse the lignin and ash product with acetone, dry at 105°C for 2 hr, cool in a desiccator and weigh (W4).

6 Ignite sinter at 550°C for 2 hr, cool in a desiccator and re-weigh (W5).

Calculations

Lignin (%) = {(W4 - W5) / W1} x 100 x {100 / 100 - % water content}

Cellulose (%) = {(W3 - W4) / W1} x 100 x {100 / 100 - % water content}

References

Allen, S.E. (ed.) (1974) *Chemical Analysis of Ecological Materials.* Blackwell Scientific Publications, Oxford, pp.252
Browning, B.L. (1967) *Methods of Wood Chemistry* Vol. 2. John Wiley and Sons, New York.
Clancey, M.J. and Wilson, R.K. (1966) Development and application of a new chemical method for predicting the digestibility and intake of herbage samples. *Proceedings 10th International Grassland Congress*, pp.445-453.
Morrison, I.M. (1972) A semi-micro method for the determination of lignin and its use in predicting the digestibility of forage crops. *Journal of the Science of Food and Agriculture* **23**, 455-465.
Rowland, A.P. and Roberts, J.D. (1994) Lignin and cellulose fractionation in decomposition studies using acid-detergent fibre methods. *Communications in Soil Science and Plant Analysis* **25,** 269-277.

Southgate, D.A.T. (1967) Determination of carbohydrates in food. II. Unavailable carbohydrates. *Journal of the Science of Food and Agriculture* **20**, 331-335.

Van Soest, P.J. (1963) Use of detergents in the analysis of fibrous feeds. II: A rapid method for the determination of fibre and lignin. *Journal of the Association Off Agricultural Chemists* **46,** 829-835.

Chapter 7 SOIL PHYSICAL ANALYSES

7.1 MECHANICAL ANALYSIS (TEXTURE)

Reagents

Amyl alcohol
Hydrogen peroxide, 30 % (100 volumes)
Sodium hexametaphosphate

Procedure

1 Weigh 50 ± 0.5 g soil into a 500 ml heat resistant (105°C) screw lid bottle calibrated at 250 ml.

2 Add 125 ml water and swirl the mixture to wet the soil thoroughly.

Note: Steps 3, 4 and 5 can be omitted if the soil is very low in organic carbon, i.e. < 1%.

3 Add 20 ml 30% hydrogen peroxide and gently swirl the bottle. Add 1 or more (as necessary) drops of amyl alcohol and gently swirl to minimise foaming.

4 When the reaction has subsided, heat the bottle in a boiling water bath to complete the reaction. Again add amyl alcohol drop-wise to contain the bubbles.

5 When the reaction has subsided, remove the bottle from the water bath and allow to cool.

6 Add 2.0 g sodium hexametaphosphate and make up to the 250 ml mark with water.

7 Shake end over end for about 18 hr, i.e. overnight.

8 Transfer the contents to a 1000 ml sedimentation cylinder, add the water washings, and make up to the 1000 ml mark with water.

9 If the fluid temperature is likely to fluctuate by ± 3°C during the day, place the cylinder in a water bath so as to maintain a constant temperature.

10 Make up a blank cylinder of 2.0 g sodium hexametaphosphate dissolved in water, made up to the mark with water, and place this in the same tank. Allow the cylinders to equilibrate for 30 min.

11 Mix the sample cylinder vigorously with a plunger, and start the stop-watch the moment the plunger is removed.

12 Record the Bouyoucos hydrometer readings at 40 sec and 5 hr, the 5 hr blank cylinder hydrometer reading and the tank temperature. (If no hydrometer is available, pipette 25 ml from a depth of 10 cm at the 5 hr time into a pre-weighed dish, evaporate to dryness and weigh; this will give clay.)

Note: For fine resolution determination of silt and clay (e.g. for investigating faunal effects on soil texture), use the pipette method as decribed in Gee and Bauder (1986).

Calculations

40 sec (corr) = 2(40 sec reading - 40 sec blank + T)
5 hr (corr) = 2(5 hr reading - 5 hr blank + T)

where
T = temperature corrections: For every °C above 20°C (d), T = 0.3 x d; for every °C below 20°C (d) T = -0.3 x d.

% sand = 100 - 40 sec (corr)
% silt = 40 sec (corr) - 5 hr (corr) % clay = 5 hr (corr) where:

for: sand = 2 mm - 0.06 mm; silt = 0.06 mm - 0.002 mm; clay = <0.002 mm

Textural classification according to USDA should be followed (see Figure 7.1).

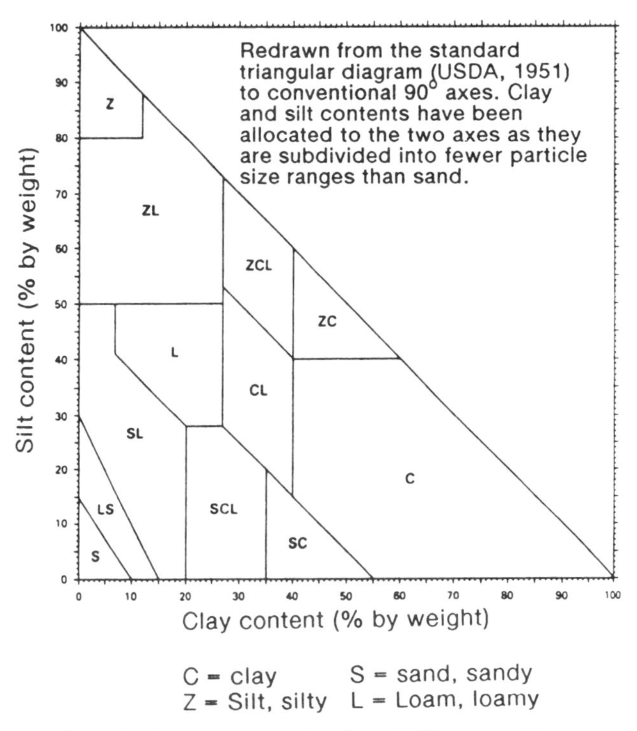

Figure 7.1 Two dimensional chart for assigning USDA soil texture classification (from Booker Tropical Soil Manual, with permission).

Reference

Gee, G.W. and Bauder, J.W. (1986) Particle-size analysis. In: Klute, A. (ed) *Methods of Soil Analysis, Part 1.* Second Edition. American Society of Agronomy, Inc., Madison.

7.2 BULK DENSITY

The ideal soil water content for measuring bulk density is field capacity, although minor deviations from field capacity will not result in significant errors. Determinations, however, should not be carried out when the soil is very dry.

7.2.1 Method for non-stony soils

Procedure

1 Remove 1-2 cm of surface soil from the spot where samples will be taken, and level the spot.

2 Drive a 5 cm diameter thin-sheet metal tube of known weight (W1) and volume (V) 5 cm into the soil surface.

3 Excavate the soil from around the tube and cut the soil beneath the tube bottom.

4 Trim excess soil from the tube ends.

5 Dry at 105°C for 2 days, and weigh (W2).

Calculation

Bulk density (g/cm^3) = (W2 - W1) / V

Reference

Blake, G.R. and Hartge, K.H. (1982) Bulk Density. In: A. Klute (Ed) *Methods of Soil Analysis, Part 1.* Second Edition. American Society of Agronomy, Inc., Madison.
Lutz, J.F. (1947) Apparatus for collecting undisturbed soil samples. *Soil Science* **64**, 399-401.
Uhland, R.E. (1950) Physical properties of soil modified by crops and Management. *Soil Science Society of America Proceedings* **14**, 361-366.

7.2.2 Methods for stony soils

Procedure

1 Excavate an intact clod of soil (about fist size).

2 Air-dry the clod, tie a thin thread around it and weigh it (W_s).

3 Dip it briefly in melted paraffin wax (60°C) to waterproof it.

4 Weigh the coated clod (W_{sp}) and calculate the weight of the paraffin coating ($W_p = W_{sp} - W_s$).

5 Suspend the clod from the balance arm and submerge it completely in a beaker of water. Record the weight (W_{spw}). If it leaks air, discard it.

6 Break open the clod, take a subsample of the soil, weigh it, oven-dry it, and re-weigh.

Calculation

Correct W_s to its oven-dry mass, W_{dry}:

$W_{dry} = W_s$ x (weight of subsample after drying / weight of subsample before drying)

Bulk density = $W_{dry} / [\{(W_{sp} - W_{spw})/\rho_w\} - (W_p/\rho_p)]$

where
ρ_w = density of water at temperature of determination (1.0)
ρ_p = density of paraffin wax (approximately 0.9)

If soil conditions are such that the above method is impractical, an infill method may be used:

Procedure

1 Dig a hole approximately 10 cm x 10 cm x 10 cm. Dry all the soil removed at 105°C for 24 hr and weigh (W).

2 Fill the hole with dry coarse sand from a known volume of sand. Ensure the sand surface is level with the surrounding soil surface. Record the volume of sand remaining and hence calculate the volume used to fill the hole (V).

Alternatively, press ultra-thin plastic food wrap around the sides of the hole, and fill with water from a known volume of water. Record the volume of water remaining and hence calculate the volume used to fill the hole (V).

Calculation

Bulk density (g/cm^3) = W / V

7.2.3 Method for shrink-swell soils

In shrink-swell soils the wet bulk density and the dry bulk density differ. It is difficult to measure when the soil is either very wet or very dry. The minimum bulk density (which occurs when the soil is fully wet) can be estimated by assuming that the particle density is 2.65 g/cm and air content when field saturated is 0.03 cm^3/cm^3 (Shaw and Yule, 1978):

$$\text{Bulk density}_{max} = 1 / (W_{max} + 0.4046)$$

where W_{max} is the maximum gravimetric water content of the horizon in question (the field saturated water content).

Reference

Shaw, R.J. and Yule, D.F. (1978) *The Assessment of Soils for Irrigation, Emerald, Queensland.* Technical Report 13, Agricultural Chemistry Branch, Queensland Dept. of Primary Industries, Australia.

7.3 FIELD CAPACITY

"Field capacity" is an arbitrary point on a curve. It is defined as the maximum amount of water the soil can hold after it has freely drained. In practice, as this defined point is hard to ascertain, field capacity is estimated after the soil has drained for 2 - 3 days following saturation, but before evapotranspiration has depleted the water store.

Procedure

1. Build an earth wall around a 1 m x 1 m vegetation-free area, and fill with tap water.
2. Refill as necessary so that approximately 50 cm (500 litre) of water has soaked into the soil.
3. Cover the area with a plastic sheet to prevent evaporation and leave for 3 days.
4. Bulk 5 replicated 0-10 cm soil samples from near the centre of the area.
5. Put about 250 g of the wet soil in a container of known weight (W1), weigh (W2), then dry at 105°C for 48 hr.
6. Weigh the dry soil again in the container (W3).

Calculation

Gravimetric soil water content at field capacity (%) = {(W2 - W3) / (W3 - W1)} x 100

Volumetric soil water content at field capacity (%) =
Gravimetric water content at field capacity x bulk density

7.4 LOWER LIMIT OF PLANT AVAILABLE WATER

This value is sometimes known as wilting point and is often equated to the soil water content at -1.5 MPa (-15 bar) water potential. A more biologically valid way of measuring it is in the field under vegetation of the type being studied. A fully developed stand of plants is deprived of water (usually by waiting for a dry period, if a rain-exclusion shelter is not available) and the water content of the soil is determined at its minimum value by taking gravimetric samples at several depths down to the bottom of the effective rooting depth or by using a calibrated neutron probe.

In permanently wet climates this is not an acceptable practical method, and the -1.5 MPa water content has to be measured.

7.4.1 The -1.5 MPa water content using a pressure chamber apparatus

Procedure

1 Distribute rubber sample rings or metal rings with cheesecloth fastened to one end with a rubber band around a pre-soaked -1.5 MPa ceramic plate.

2 Fill each ring with soil. Do not compress or pack the soil into the ring. Prepare triplicate samples.

3 Place the plate in a large tray and slowly add tap water until the water is about half way to the top of the sample rings. Soak the samples overnight.

4 Seal the outflow tube on the ceramic plate with a clamp. Carefully drain excess fluid out of the tray. A syringe or siphon works well.

5 Place the plate (with samples) in the pressure chamber. Connect the outflow tube of the ceramic plate to the fitting on the inside of the chamber. Connect another tube to the fitting on the outside of the chamber, and place the free end of the tube in a beaker of water. Unclamp the outflow tube so that water may flow freely from the ceramic plate to the beaker on the outside of the chamber.

6 Place a damp cloth over the samples in the chamber to maintain high humidity while the samples are equilibrating. Close the chamber, tighten the wing nuts, and slowly apply air pressure to the chamber until 1.5 MPa is reached.

7 Allow the samples to equilibrate for 2 to 4 days. The longer time is for soils with high clay contents.

8 Before releasing air pressure, clamp the outflow tube so that water cannot re-enter the ceramic plate.

9 Release pressure slowly. Open the chamber and remove the samples. Determine the gravimetric water content as in Section 6.1.

Calculation

$$\text{Lower limit of plant available water (\%)} = \{(W2 - W3) / (W3 - W1)\} \times 100$$

where
W1 = weight of container (g)
W2 = weight of container + wet soil (g)
W3 = weight of container + oven-dry soil (g)

Reference

Klute, A. (1986) Water Retention: Laboratory Methods. In: A. Klute (ed). *Methods of Soil Analysis, Part 1.* Second Edition. American Society of Agronomy, Inc., Madison.

7.4.2 Soil water potential by the filter paper method

The water potential of relatively dry soil can be estimated by allowing it to equilibrate with a piece of Whatman 42 filter paper, and then determining the water content of the filter paper. There is a consistent relationship between filter paper water content and equilibration potential, over the range -66 to -4 500 kPa. To determine the -1.5 MPa water content of soil using this cheap, low technology method, allow several soil samples of about 200 g each to dry for varying lengths of time, and determine their water contents (6.1) and water potential (see below). Plot the water content against the logarithm of the absolute water potential, and interpolate the water content at -1.5 MPa.

Procedure

1 Seal a sample of about 200 g into a air-tight container with two 70 mm diameter disks of Whatman 42 filter on top of the soil.

2 Leave the container in a dark, even-temperatured place (an insulated picnic box is good) for 72 hours.

3 Open the container, and quickly weigh the upper filter paper disk to 0.0001 g (W1). Dry the filter paper at 105°C for 2 hr, and reweigh it to the same accuracy (W2).

Calculation

The filter paper water content

$$\%WC = (W1-W2)/W2 \times 100$$

The equilibrium water potential (WP, kPa)

$$WP = -\exp(-0.128 \times \%WC + 11.424) \quad \text{if } \%WC < 56.2$$
$$WP = -\exp(-0.019 \times \%WC + 5.282) \quad \text{if } \%WC > 56.2$$

Reference

Savage, M.J., Khuvutlu, I.N. and Bohne, H. (1992) Estimating water potential of porous media using filter paper. *South African Journal of Science* **88**, 269-274.

7.5 PLANT AVAILABLE WATER CAPACITY (PAWC)

The amount of water which a given soil horizon can store for plant use is estimated from the difference between the field capacity and lower limit of plant available water for that horizon. It is expressed as an equivalent depth of water (mm), and is calculated using the values for field capacity (Section 7.3) and the lower limit of plant available water (Section 7.4):

$$PAWC = \{(FC - LL) / 100\} \times D_{soil} \times Z$$

where

FC = gravimetric field capacity (%)
LL = gravimetric lower limit of plant available water (%)
D_{soil} = bulk density of the horizon (g/cm^3)
Z = thickness of the horizon (mm)

The total PAWC is the sum of the PAWC of all the horizons down to the effective rooting depth.

Reference

Shaw, R.J. and Yule, D.F. (1978) *The Assessment of Soils for Irrigation, Emerald, Queensland.* Technical Report 13, Agricultural Chemistry Branch, Queensland Dept. of Primary Industries, Australia.

7.6 INFILTRATION

Infiltration refers to the vertical intake of water to the soil surface. The most commonly used method to determine infiltration is to flood an area contained within a bund, and to record the water level over time (ponded method). Infiltration rates measured under a sprinkler or under natural rainfall may not be the same.

The measured infiltration rate is markedly affected by cylinder diameter, the rate measured being lower for larger diameters because of the reduced effect of lateral flow. Better estimates are therefore obtained if two concentric rings are used, the water level being maintained in both, while measurements are made in the inner; this compensates in part for lateral flow. An excellent general reference can be found in The Booker Tropical Soil Manual (Landon, 1991). The procedure described in Section 7.6.1 is primarily used for site characterisation, notably for irrigation surveys. A single ring of diameter 25 - 30 cm (Section 7.6.2) may be used to monitor treatment effects in field experiments thereby minimizing site disturbance, and using much less water; replication is therefore easier.

7.6.1 Double ring method

Procedure

1 Remove surface litter from an area 1.5 m x 1.5 m.

2 Pre-wet the area by soaking for a few hours using an earth bund to contain the water.

3 Vertically drive the metal rings (inner 30 cm diameter, outer 60 cm diameter, heights 50 cm) about 15 cm into the wet soil so that the smaller ring is centred in the larger. (This may be facilitated by cutting the soil with a knife and sealing the cylinder in the soil with bentonite.) Lay sacking on the soil surface within the rings to minimize disturbance from adding water.

4 Fill both cylinders to about 15 cm, record the time and the distance from the water level in the inner cylinder to the inner cylinder top.

5 Measure the water level at 1, 5, 10, 20, 30, 45, 60, 90 and 120 min, or more frequently if infiltration is rapid.

6 Refill the cylinder when the level has dropped to about 10 cm, noting the water level before and after refilling on each occasion. Try to maintain equivalent water levels in the inner and outer rings.

7 Continue the measurements until a steady state has been reached, e.g. 1 - 4 hr.

8 Construct a table of results to show time intervals (min), cumulative time (min), intake (cm), cumulative intake (cm) and infiltration rate (cm/hr).

9 Plot infiltration rate (cm/hr) against cumulative time (min).

The basic infiltration rate (cm/min) is that when steady state (i.e. a straight line on the graph) is attained.

Reference

Anon (1991) Soil Physics In: Landon, J.R. (ed.), *Booker Tropical Soil Manual.* Longman, Harlow, UK. pp. 59-71, and pp. 213-217.

7.6.2 Single ring method

Procedure

1 Remove surface litter from a 50 cm x 50 cm area of bare soil.

2 Vertically plastic or metal rings (diameter 25 - 30 cm, length 15 cm) 2 - 3 cm into the soil. Press soil around the base to minimise water leakage, and cover the soil surface in the ring with sacking.

3 Fill the cylinder to about 10 cm deep, record the time, and measure the water depth.

4 Continue as described in Section 7.6.1, from step 4.

Reference

Berryman C. and Trafford, B.D. (1974) *Infiltration rates and hydraulic conductivity.* Field Drainage Experimental Unit Technical Bulletin, MAFF, London. 16pp.

7.7 POROSITY

Soil pores are classified by size. Macro-pores (diameter > 0.1 mm) are chiefly responsible for aeration and gravity flow; meso-pores (diameter 30 - 100 μm) conduct water by rapid capillary flow; micro-pores (diameter < 30 μm) are responsible for water retention and slow capillary flow.

7.7.1 Total porosity

Total porosity (the sum of all pore size volumes) is calculated from the dry bulk density (cf. Section 7.2) and the particle density (assumed to be 2.65 g/cm^3 for most mineral soils):

Total porosity (%) = {1 - (bulk density / particle density)} x 100

7.7.2 Pore size distribution

Pore size distribution is estimated on an undisturbed soil core. The water content of the soil after all pores greater than a given size have been drained is determined volumetrically and subtracted from the total porosity. This provides an estimate of the volume of pores above a specified minimum pore diameter. Macro-pores and the larger meso-pores (> 50 μm) can be drained by suction applied to the core using a tension table. To calculate the water tension to be applied:

Tension (cm water) = 2908/minimum pore diameter (μm)

e.g. all pores >300 μm will be drained at 2908/300 = 9.7 cm water tension. The tension table has a range to measure water release of pores down to about 50 μm.

Procedure for making a suitable tension table

1 Drill a hole in the centre of a plastic tray or sheet (0.5-1 m diameter) (table) and attach a fitting for the connection of flexible PVC or polythene tubing.

2 Place a plastic screen on the table which is about 10 cm smaller than the diameter of the table.

3 Place blotting paper on top of the screen. The paper should be the same size as the table.

4 Attach tubing to the fitting on the table, and place the free end of the tubing in a flask with an overflow.

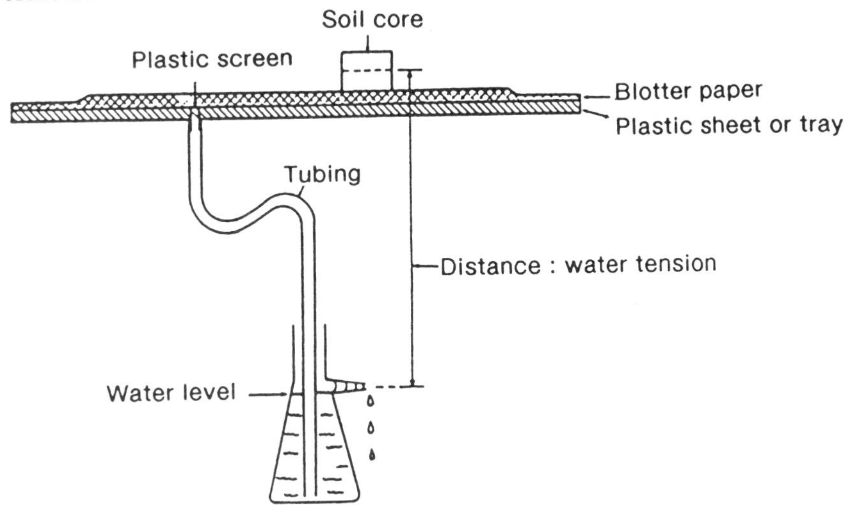

Figure 7.2 Tension table apparatus.

Procedure

Note: Care in sampling is extremely important, because disturbances to the soil core may greatly alter porosity measurements.

1 Sample when the soil is fairly moist (about field capacity). Use a core sampler of known weight (W1) with a bevelled cutting edge to minimise soil disturbance. Cores should be about 2.5 cm high and 5 cm in diameter.

2 Obtain four to six cores. (If the top soil is very loose, sample a few cm below the surface.)

3 Trim the excess soil from the cores and fasten muslin to one end of each core with a rubber band.

4 Place the cores in a pan and slowly add water until the level is just below the top of the cores. Leave to soak overnight.

To prepare the tension table (see Figure 7.2):

1 Fill the overflow flask with water and put the PVC tube into it. Raise the flask above the table and fill the tube with water.

2 The region between the blotter paper and table must be filled with water, and there should be as few bubbles beneath the blotter paper and in the tubing as possible. This will take some practice to achieve.

3 Using a rolling pin or your fingers, press the blotter paper against the table so that a seal is formed when the flask is lowered beneath the table. Air should not leak into the area below the blotter paper or the tubing at any time. If air does enter the system, the measurements will be inaccurate.

4 Lower the flask so that the distance from the overflow tube to the centre of the sample is equal to the amount of tension needed to drain pores above the desired minimum, i.e. for 300 μm this distance is 9.7 cm.

5 Carefully drain the tray with the samples in it using a siphon.

6 Using care, move the samples onto the tension table (muslin side down). Cover the table with plastic or a damp cloth to maintain high humidity.

7 Leave the samples on the tension table for two to three days. Make sure that no air has entered the tension table system, and that the water level in the flask is always at the level of the overflow.

8 Remove the cores, measure their height and diameter carefully and calculate the sample volume (V) in cm^3 ($\pi r^2 h$). This measurement is very important for calculating porosity, and should be made using vernier or dial callipers if they are available.

9 Weigh the samples, being careful to not lose any water or soil while handling (W2). Dry the samples in a 105°C oven for 48 hr and re-weigh them (W3).

Calculations

Volumetric water content (%) = {(W2 - W3) / V} x 100

Macro porosity (%) = total porosity - volumetric water content

Note: A volume for any given size range can be measured by equilibrating cores at tensions corresponding to the upper and lower pore sizes of interest and using the difference in water content at each tension to calculate the volume of pores between those pore sizes.

References

Bouma, J. (1981) Comment on "Micro-, Meso-, and Macroporosity of Soil". *Soil Science Society of America Journal* **45**, 1244-1245.

Danielson, R.E. and Sutherland, P.L. (1986) Porosity. In: Klute, A. (ed.), *Methods of Soil Analysis, Part 1.* Second Edition. American Society of Agronomy, Inc., Madison.

Luxmoore, R.J. (1981) Micro-, Meso-, and Macroporosity of Soil. Letter to the Editor. *Soil Science Society of America Journal* **45**, 671-672.

Appendix A EXAMPLE OF RAPID RURAL APPRAISAL (RRA) INTERVIEW

Lenora Bohren

A.1 INTRODUCTION

According to Grandstaff *et al.* (1987), Rapid Rural Appraisal (RRA) is a new research paradigm which views the world as a highly interactive set of variables which are rapidly changing and subject to a high degree of uncertainty. The paradigm is based on the concept of adaptive change involving feedback systems utilizing the perceptions of differing populations. RRA is used to:

1 explore, identify, and diagnose rural situations, problems, or issues;
2 design, implement, monitor, and evaluate programs, projects, and development;
3 help develop, extend, and transfer technology;
4 formulate policy and assist decision making;
5 respond to emergencies and disasters; and
6 improve, supplement, or complement other research.

A.2 PHASE I: PREPARATORY WORK

A Selection of an interdisciplinary team according to the research topic (Grandstaff *et al.*, 1987). Teams can be paired in terms of their discipline as in the Sondeo Approach (Hildebrand, 1981) and/or paired with local researchers as suggested by Chambers (1991). Smaller teams are often preferred (Beebe, 1985).

B Team examination of relevant background information gathered from published and unpublished reports such as archives, annual reports, census reports, maps, aerial photographs, environmental and social data, etc. on:

1. Natural circumstances such as rainfall, amount and reliability; seasonal temperatures; soil characteristics, topography, such as slope and depth, and hydrology; and weeds, pest and disease incidence as a source of uncertainty of crop output (Grandstaff *et al.*, 1987; Beebe, 1985; CIMMYT, 1985; Doorman, 1991).

2. Institutional circumstances such as types of marketing and supply channels; types and reliability of food distribution channels; extension and credit programs, types of programs and numbers and types of participants; land tenure arrangements; farmer groups, voluntary, organized, official or unofficial, and planned actual functions; and other infrastructure considerations (Beebe, 1985; CIMMYT, 1985; Grandstaff *et al.*, 1987).

3. Socio-economic circumstances such as population numbers, density and settlement patterns; available area and production figures; marketed products, volume and trends for outputs sold and inputs purchased; food purchased; relative volume and trends and prices over the years and between seasons, prices, and marketing margins; and available information (most of which will come from the in-depth interviews and observation) on goals, beliefs, attitudes, and social obligations (Beebe, 1985; CIMMYT, 1985; Grandstaff *et al.*, 1987).

C Team discussion of research topic and background information aimed at developing preliminary hypotheses (if appropriate), interview topics, methods, tools, and techniques related to the research topic (Grandstaff *et al.*, 1987).

D Team consultation with persons knowledgeable of the topic or geographic area if possible (Grandstaff *et al.*, 1987).

E Team preliminary field visit to familiarize team with area, to discuss the project with local officials, to pretest the interview checklist and identify key indicators, and to select, if possible, local researchers (Beebe, 1985; Grandstaff *et al.*, 1987; Chambers, 1991). Examples of checklists are in Beebe (1985), Raintree (1987), Rocheleau *et al.* (1988) and IBSRAM (1988).

F Team discussion and planning of research sites, logistics and protocols for the fieldwork (Grandstaff *et al.*, 1987; Raintree, 1987).

A.3 PHASE II: FIELDWORK

A Primary data collection by research team from interviews and observation. The RRA depends on walking, seeing, and asking.

1. Interviews are conducted using a checklist as a flexible guide and should be carefully timed so as not to interfere with work schedules. The use of semi-structured interviews (checklists) are essential for obtaining information such as indigenous technical knowledge (ITK). Follow up interviews are conducted as needed. The number and length of the interviews depend on the homogeneity of the population (Beebe, 1985). The persons interviewed are:

a) Individuals representative of a cross section of the expected target population. A cross section of farmers would include farmer leaders, women farmers, major cropping farmers, poor farmers, rich farmers, traditional farmers, and innovative farmers. Rhodes (1982) stresses the importance of including all farm members who are involved in agricultural decision making, especially those who have migrated off the farm or are working in off farm occupations.

b) Key informants can be farmers and non-farmers. They should include those who have experience in the system beyond their own participation as farmers such as bankers, landlords, ministry officials, merchants, middlemen, extension agents, buyers of agricultural products, and suppliers of inputs (Beebe, 1985).

c) Groups representing the variability within the community. Chambers (1991) suggests a chain of interviews from experts to novices in a process such as farming. Group interviews gather information such as natural history, local histories and sensitive information such as land holdings. Group interviews can be self correcting, participants often correct each other's information. However, in group interviews, it is important to observe behaviour to help distinguish between real and ideal practices since the presence of others may influence answers. Some issues may need to be confirmed by individual interviews. Suggested group interview techniques are the "informal Delphi" technique or focus group techniques (Honadle, 1979; Beebe, 1985; Bryant and Bailey, 1991).

2. Observation includes direct observation of what people are doing, the use of photographs, actual participation, observing several farmers, etc. It is important to have multiple checks on information and take good, detailed field notes. Check lists are often used. It is important to look for patterns such as crop production, land use, and farmer behaviour. This can be done by agroecological transects and field plotting by air or foot, taking pictures, and direct observation of key indicators. Key indicators can be soil colour as an indicator of particle size and distribution, fertility, and drainage properties; birth weight of children as an indicator of health and nutrition; housing as an indicator of poverty or prosperity; soap inventories as an indicator of purchasing power; the appearance of new items in adjacent areas as an indicator of the trickle down of benefits; transfers and turnovers as an indicator of organizational capability (Chambers, 1980); and/or others suggested by the local participants.

B Single or multiple visits to a particular village or visits to a number of villages (Beebe, 1985). Also visits outside study area are important in understanding the dynamics of the study area (Grandstaff *et al.*, 1987).

C Continuous team meetings and discussions (Grandstaff *et al.*, 1987).

D Field research should last anywhere from four days to three weeks depending on funding and time restraints (Grandstaff *et al.*, 1987).

E One or several field study periods depending on the nature of the problem. Follow-up visits are recommended (Beebe, 1985; Grandstaff *et al.*, 1987).

A.4 PHASE III: COMPLETING THE RRA

A Team discussion and analysis immediately following fieldwork.

B Results formalized and recorded. The report is jointly written by the team or by one individual in consult with the rest of the team. The report should include the results, methodology and recommendations. Data collection checklists should be attached to the report. Audiovisual aids should be used where appropriate (Beebe, 1985; Rhodes, 1982; Grandstaff *et al.*, 1987).

C Copies of the report sent to all interested agencies, offices, and individuals. Publication should be done where appropriate (Grandstaff *et al.*, 1987).

D Learning should not stop when report is written. RRAs can often result in future research projects and need to be factored into decision making (Grandstaff *et al.*, 1987; Beebe, 1985).

E Follow-up and evaluation of how well the target population is adapting to the new innovation is crucial. This can be done in the iterative process or by a follow up study if funds are sufficient.

References

Beebe, J. (1985) *Rapid Rural Appraisal: The Critical First Step in Farming Systems Approach to Research.* Farming Systems Support Project, Networking Paper No, 5, USAID, Washington, D.C.

Bryant, C.A. and Bailey, D.F.C. (1991) The Use of Focus Group Research in Program Development. In: van Willigen, J. and Finan, T.L. (eds.), *Soundings: Rapid and Reliable Methods for Practising Anthropologists, NAPA Bulletin* **10**, 24-39, American Anthropological Association, Washington, D.C.

Chambers, R. (1980) *Short Cut Methods in Information Gathering for Rural Development Projects.* Paper for World Bank Agricultural Sector Symposia, January.

Chambers, R. (1991) In: Cernea, M.M. (ed.), *Putting People First: Sociological Variables in Rural Development*, 2nd Edition. Oxford University Press, New York, pp.515-534.

CIMMYT (1985) *Teaching Notes on the Diagnostic Phase of OFR/FSR Concepts, Principles and Procedures.* An occasional series of papers and notes on methodologies and procedures useful in farming systems research in the economic interpretation of agricultural experiments, No. 14, November. CIMMYT East African Economics Programme, Nairobi, Kenya.

Doorman, F. (1991) A framework for the rapid appraisal of factors that influence adoption and impact of new agricultural technology. *Human Organization* **50**, 235-244.

Grandstaff, S., Grandstaff, T. and Lovelace, G. (1987) *Proceedings of the 1985 International Conference on Rapid Rural Appraisal.* Khon Kaen University, Khon Kaen, Thailand: Rural Systems Research and Farming Systems Research Project.

Hildebrand, P.E. (1981) Combining Disciplines in Rapid Appraisal: The Sondeo Approach. *Agricultural Administration* **8**, 423-432.

Honadel, G. (1979) *Organization and Administration of Integrated Rural Development.* Working Paper No. 1. Rapid Reconnaissance Approaches to Organizational Analysis for Development Administration. Development Alternatives, Inc. Washington, D.C.

IBSRAM (1988) Methodological Guidelines for IBSRAM Soil Management Networks. Second draft, April. IBSRAM, Bangkok, Thailand.

Raintree, J.B. (1987) *D and D User's Manual. An Introduction to Agroforestry Diagnosis and Design.* ICRAF, Nairobi, Kenya.

Rhodes, R. (1982) *The Art of the Informal Agricultural Survey.* Training Document 1982-2, Social Science Department, International Potato Centre. Lima, Peru.

Rocheleau, D., Weber, F. and Field-Juma, A. (1988) *In Dryland Africa.* ICRAF, Nairobi, Kenya.

Appendix B THE RHIZOSPHERE

Glynn Bowen

B.1 INTRODUCTION

This appendix gives a general overview of the root environment (the rhizosphere), its physical, chemical and biological nature, and the processes occurring there. A fairly detailed entry into the literature is given by Bowen and Rovira (1976), Foster and Bowen (1982), Rovira *et al.* (1983), Marschner (1986) and Kesster and Cregan (1991). As well as the non-infecting organisms which grow in the rhizosphere, most organisms infecting plant roots, be they symbionts or pathogens have a rhizosphere phase before infection occurs and/or during spread of the organism to other infection sites. The growth/survival of symbionts in the rhizosphere can be critical to symbioses. Nevertheless, these are not singled out for special discussion in this appendix. Mycorrhizas are discussed in detail in Appendix C.

The rhizosphere refers to the zone of soil adjacent to roots. This has been further subdivided by some to the "outer rhizosphere", the "inner rhizosphere" and the "rhizoplane" i.e. the root surface itself. The extent of the rhizosphere is quite ill-defined and variable depending on the soil conditions affecting diffusion of solutes or volatiles from the root (and their biological activity); - it can be from 1 mm to approximately 10 mm (e.g. with volatiles). The inner rhizosphere, the zone of most intensive change, is generally considered to extend 15-20 μm from the root.

B.2 THE RHIZOSPHERE ENVIRONMENT

Losses of organic substrates from roots to the soil can be substantial - usually between 10% and 20% of plant production in non-sterile soil (see Rovira *et al.*, 1983). These large losses, which started emerging from studies in the mid 1970s, are 10-100 times those indicated by the earliest studies of root exudation. There are several reasons for this discrepancy: losses from solution culture (early studies) are considerably less than into a solid medium such as soil (the presence of soil micro-organisms approximately doubles these losses) and studies in sterile solution culture often use only very young plants (losses increase as roots age because of an increase in root senescence).

Much of the early literature refers to losses of organics from roots as "root exudates", but it is now clear that "losses" can be from several sources (Rovira *et al.*, 1979). Loss of soluble compounds from young, intact cells have been distinguished as:

1 "exudates" (leaking passively from cells). These low molecular weight substances are readily available to micro-organisms for energy and include sugars, organic acids, amino acids and several other compounds. The ratio of each varies with

environmental factors which affect the composition of the cytoplasm (see Bowen and Theodorou, 1973, Graham *et al.*, 1982); often sugars and organic acids far exceed amino acids; and

2 "secretions" (compounds of low and high molecular weight lost by active movement out of the cell or "secretion"). Roots are usually surrounded by mucilages (high molecular weight polysaccharides) of both plant and bacterial origin (collectively referred to as "mucigel"). The plant derived polysaccharides are probably produced by Golgi bodies and secreted by root cap cells and epidermal cells near the apex. There is little doubt that this is secretion in its true sense. Similarly H^+ extrusion pumps occur in root cells. However, with low molecular weight compounds there are almost no good data on whether they are leaked along a concentration gradient (passive loss, "exudation") or actively expelled or "secreted". Because the composition of the solutes lost from roots is often similar to that of the roots themselves, passive loss (exudation) may be the major mechanism of loss of low molecular weight compounds from intact cells.

Cells which are senescing, however, may be a major source of organic loss (perhaps <u>the</u> major source once a cell matures). Such losses have been identified as "lysates". They increase as the root matures and ages, e.g. from the cortex and sloughed root cells. There is some evidence that reports of extensive death of cortical cells of young roots (e.g. Henry and Deacon, 1981) may be due to staining artifacts (Wenzel and McCully, 1991). With roots growing in non-sterilised soils, a number of the cells only a centimetre or two from the apex, are empty and heavily colonised by bacteria; it may be that some of these cells were killed prematurely by sub-clinical pathogens. The process of ageing and death of cells increases with age of the root and is probably accelerated in the presence of soil organisms (including sub-clinical pathogens). For example Martin and Kemp (1985) found when wheat was grown in fumigated soil and given a pulse of $^{14}CO_2$ at the early tillering stage, some 14.6% of the label appeared as CO_2 (from root respiration and respiration of micro- organisms growing on root substrate) in soil over 14 days but 26.7% was lost as CO_2 from unfumigated soil.

Sterile roots senesce with time, releasing "lysates" and the important fact is that micro-organisms may change the time course of this dramatically, which in turn could have follow-on consequences for soil microbial processes. Even in the absence of micro-organisms, however, there is a large throughput of organic matter from roots into soil even quite early in the plant's life.

How important quantitatively is carbon loss from intact cells, compared with carbon losses via respiration and via "lysates"? The most appropriate methods to find out involve $^{14}CO_2$ labelling in closed systems and recovering ^{14}C from shoots, roots, soil and periodic trapping of $^{14}CO_2$ flushed from the soil (a measure of root respiration and microbial respiration from root derived substrates) (Barber and Martin, 1976). The soil ^{14}C can be separated into water soluble fractions (probably uncharged solutes and some cations) and water insoluble fractions (some high molecular weight compounds, bacterial colonies, polysaccharides, sloughed root fragments).

The data in Table B.1 from Barber and Martin (1976) indicate the relative order of various fractions during the first three weeks of a wheat plant. Note that in sterile plants only some 0.9% of ^{14}C assimilated was lost as low molecular weight compounds and may be largely exudate, i.e. slightly less than root respiration. In non-sterile plants the situation was less clear, but a considerable portion of the respired $^{14}CO_2$ (almost 8% of the total assimilate of the plant compared with 1% in sterile soil) would have arisen from increased leakage from the root and some breakdown of sloughed plant parts, partly induced by the soil microflora. There appears to be little profitability in breeding plants with less leakage of materials from intact cells, if the 1% indicated (from sterile plants) in Table B.1 is typical. The value of manipulation of the "insoluble" fraction (some 5-6%) is debatable due to lack of knowledge of its composition (polysaccharide or root cells) and the function and "need" for the polysaccharide production by the root, if any.

The high molecular weight polysaccharides produced by the plant have been partly characterised by electron-microscope studies in which various hydrolysis agents are used and the residual polysaccharide stained (see Foster and Martin, 1981; Foster and Bowen, 1982). Their functions have been speculated on, two of the most attractive being that they may act as a lubricant for the growing root through soil, but more particularly that they may serve as an effective continuous contact between soil and the root passage of nutrients and water.

Although it is generally thought that root "exudation" and therefore substrate for micro-organisms is greatest near the root tip, quite active colonisation of parts of the roots which are 10 days old can occur (Bowen and Rovira, 1973). Such studies of microbial growth on roots inoculated at various points of known temporal and physiological age will indicate the available substrate for microbial growth at any one time or place; however this will not identify the sources of the substrate - exudation or lysis.

Table B.1 Distribution and loss of photosynthetically fixed ^{14}C in wheat plants grown in sterile and non-sterile soil for 3 weeks in an atmosphere containing 0.03% $^{14}CO_2$ and with a 16 hr photoperiod (results are expressed as mean of five replicates ± standard errors).

	Distribution (% total)				
Soil condition	Roots	Shoots	Water-soluble	Soil in-soluble	Respired
Sterilised	24.45	68.18	0.88	5.45	1.02
	1.76	±2.32	±0.05	±0.56	±0.06
Non-sterile	32.13	51.85	0.62	7.71	7.73
	±2.65	±2.37	±0.15	±0.41	±0.68

B.2.1 The movement of organic substrates into soil

If low molecular substrates (solutes and volatiles) are not used by microbial growth at the root surface, they can diffuse from the root. Thus a concentration gradient away from the root occurs, the steepness of which depends on the diffusion coefficient of the substrate, and modifications caused by interacting soil factors such as tortuosity of paths through soil, charge on clay surfaces, soil moisture etc. The distribution of exudates and microbial growth with respect to time and distance from the root has been modelled by Newman and Watson (1977). Calculations of microbial growth was 234 and 23.5 $\mu g/cm^3$ soil in wet soil and 7355 and 4.4 $\mu g/cm^3$ in dry soil at the root surface and at 0.3 mm from the root surface respectively. Such modelling has many important potential uses but it must be remembered that it is a simplified version of the situation, often with many untested assumptions, and that the predictions by any model must be tested.

In general, carbon dioxide and other volatiles will diffuse through soil for greater distance than solutes. Distant from the root (e.g. 4 to 5 mm), the volatiles may not be important as substrates, but they may be important in triggering biological activity, after which growth may follow the concentration gradient to the root (reviewed by Stotzky and Schenck, 1976). Roots may also produce exudates with specific stimulative effects on particular micro-organisms, e.g. flavonoids and isoflavonoids (Bowen, 1991; Phillips *et al.*, 1991).

Such considerations as those above are important not only for defining general microbial growth around roots but also for studying the ecology and relations between soil populations and root infection of specified economically important organisms ("epidemiology"), e.g. mycorrhizal fungi or plant pathogens. With these, and other organisms which can be recognised at the root surface, questions such as the radius of influence of the root into soil can be approached by placing propagules at fixed distances from a root and subsequently examining the root surface for the presence of the organisms (see Bowen, 1979, 1980) (see also Rhizosphere Dynamics, etc., below).

B.2.2 The inorganic nature of the root surface and rhizosphere

The root surface can differ from bulk soil in pH and in concentration of anions and cations (see Nye and Tinker, 1977). Ions which are rapidly absorbed by the root, may be at very low concentration at the root surface. On the other hand, some ions can be excluded from the root and local high concentrations can accumulate and even precipitate around the root e.g. Ca^{2+} (in some cases) and iron oxides in some flooded soils. As cations are absorbed they can be replaced by H^+, and in this way there can be a reduction of rhizosphere pH from 5 to 4 with the absorption of NH_4^+ from ammonium sulphate. Conversely, rhizosphere pH can rise following absorption of anions such as nitrate. Changes in rhizosphere pH and redox potentials are discussed by Marschner (1986). The pH of the rhizosphere of legumes fixing atmospheric nitrogen can also decline (Robson, 1982). These pH changes and associated effects on solubility of other ions can have very marked selection effects on the rhizosphere microflora (Smiley, 1979; Foster and Bowen, 1982).

Present methods for analysing the inorganic composition of rhizosphere soil include:

1 autoradiography (Lewis and Quirk, 1967; Barber and Ozanne, 1970) where ions have appropriate radioactive tracers (a semi-quantitative method); and

2 more generally, more careful collection of soil adhering to the root on removal from soil (Riley and Barber, 1969; Sakar *et al.*, 1979).

B.3 THE MICROBIAL POPULATIONS ON ROOTS

B.3.1 Culture of micro-organisms

The populations of bacteria in soil closely attached to the root (the rhizosphere soil) which are isolated on general media e.g. potato dextrose agar or Czapek-Dox agar (fungi), tryptic soy agar (bacteria) (Martin, 1975) using dilution counts or most probable number methods range from 5 - 30 x 10^8/g rhizosphere soil. A comparison of these general counts (or of specific organisms) with non-rhizosphere soil leads to the classic R:S ratio, i.e. a measure of the stimulation of micro-organisms by the rhizosphere - it can range up to 100 depending on the plant/micro-organism/growth conditions (see Rovira and Davey, 1974). Counts of bacteria on the rhizoplane on a surface basis are not common but are some 0.5-4 x 10^4/mm, but these may be spread over only some 10% of the surface (see below). However the micro-organisms growing on laboratory media may only be a small percentage of the total; Louw and Webley (1959) found that the ratio of bacteria in total (direct microscopic) counts and a laboratory media varied from 1.4 to 5.1 for rhizosphere soil. A very wide range of bacteria colonise the rhizosphere (see Rovira, 1965) and there is a selective stimulation of Gram-negative organisms requiring amino acids. One major genus is *Pseudomonas*; nevertheless, it usually represents only between 5 and 10% of the total counts.

Many fungi are also isolated from the root but dilution counts on laboratory media usually reflect the relative number of propagules of each, not the biomass, unless special techniques are used (e.g. washing and fragmentation of the roots) (see Sharley and Waid, 1955); non-sporing fungi, and slow growing fungi are missed.

In short, therefore, plate counts of rhizosphere bacteria give a distorted and partial view both of the range of organisms present and the microbial biomass on the root. Such counts are useful for general comparative purposes and in particular for studying the ecology of selected organisms. Apart from issues related to overall microbial biomass (e.g. tie up of nutrients), the study of particular organisms (usually economically important ones) is the direction which will be most rewarding for understanding rhizosphere ecology and interactions. For studying particular organisms various selective media have been developed: e.g. Simon and Ridge (1974) - fluorescent *Pseudomonas*; Brisbane and Kerr (1983) - *Agrobacterium*; and Tsao (1970) - various fungi. A further development of this is the selection of an antibiotic resistant mutant (usually a "double" mutant) of a particular species so that it can easily be isolated from a mixture of the same genus e.g. strains of *Ps. fluorescens* resistant to rifampin and naladixic acid have been used (Weller and Cook, 1983).

B.3.2 Direct observation

Direct observations of the rhizosphere have given good information on the distribution of micro-organisms; with histochemical techniques, especially at the transmission electron microscope level, insight into the physical and chemical nature of the root soil interface has been obtained. With the light microscope, staining of the rhizoplane microflora (e.g. Jones and Mollison, 1948), followed by appropriate analysis derived from higher plant ecology, e.g. quadrat counts, have shown that often only some 10% of the root surface is occupied by micro-organisms and that their distribution is non-random (see Bowen and Rovira, 1976; Rovira *et al.*, 1983). The sources of this non-randomness are twofold:

a) despite the high numbers of bacteria in soil, there is considerable spatial heterogeneity of bacterial distribution (they are usually associated with organic matter), thus the root growing through soil is not uniformly in contact with bacteria; and

b) although all parts of a root surface lose exudates, the longitudinal junctions of adjacent cells are major sites of exudation (see Bowen, 1979); these are often (not always) the preferred sites of growth and migration.

Fluorescence microscopy to distinguish dead from living organisms (Trolldenier, 1965), and the use of immunofluorescence and associate techniques such as the use of monoclonal antibody techniques to distinguish particular organisms (e.g. Malajczuk *et al.*, 1975; Schmidt, 1991), have also been powerful techniques.

Scanning and transmission electron microscopy have made important contributions to our understanding of spatial distribution of rhizosphere micro-organisms and their environment (for details, see Foster and Bowen, 1982; Foster *et al.*, 1983; Rovira *et al.*, 1983). Physiologically, the mucigels of root and microbial origin and localisation of enzymes have been a major interest. Electron probe microanalysis will add a further dimension to rhizosphere studies. Electron microscopy has also shown (i) a great diversity of bacterial forms (often not seen in rich laboratory media), many of which may be restricted to a specialised habit in grooves between epidermal cells; (ii) a ten-fold decline in microbial numbers between the root surface and 15-20 μm away (the outer rhizosphere) (Foster and Rovira, 1978); (iii) that rhizosphere bacteria are often filled with storage materials such as polyhydroxybutyrate, polysaccharide granules and polyphosphate granules; and (iv) that the outer cortex of apparently healthy roots may be inhabited by high numbers of microorganisms (the "endorhizosphere") which hasten the collapse and decay of the cortex.

It is likely that DNA probes alone, or coupled with polymerase chain reaction (PCR) amplification (Holben and Tregje, 1988; Asim *et al.*, 1990), will emerge as powerful tools in following the dynamics of rhizosphere micro-organisms without of the need to culture them. This will be particularly useful for organisms difficult to grow on laboratory media.

B.3.3 Rhizosphere dynamics, interaction and experimental approaches

The rhizosphere is a dynamic ecosystem with the same broad considerations which occur with macro-ecosystems. The same questions of growth rates, pioneers, succession and interactions (synergism, competition and antagonism) occur. A knowledge of rhizosphere ecology is important in understanding the establishment problems of inoculated beneficial micro-organisms (such as *Rhizobium*, *Frankia*, mycorrhizal fungi or bio-control organisms) and in reducing the impact of disease organisms. An appreciation of the dynamics of colonisation of and interactions in the rhizosphere can be achieved only by experimental approaches. Bowen (1979, 1980) pointed out the obvious, that once the various components of a system are identified, simple experimental approaches can be devised to test the importance of these components, e.g. spore germination, movement to the root, growth at the root, the influence of soil moisture or growth of organisms along the root and so on. Inductive (modelling) approaches to some of these features such as growth through soil have also been developed - see Gilligan (1983) (plant pathogens), and Walker and Smith (1984) (mycorrhizas).

Growth rates along roots, a corner stone of population dynamics have been studied for fungi under various conditions by inoculation of the root at known positions and measuring growth with time (e.g. Bowen and Theodorou, 1979). Growth rates for bacteria in the rhizoplane were measured for various groups of organisms by planting sterile roots 10 cm long in non-sterile soil, harvesting particular segments each at 0, 1, 2, 3 and 4 days, counting on selective media and constructing growth curves (Bowen and Rovira, 1973; Bowen, 1979). These studies indicated an initial logarithmic growth rate over 3-4 days (the slope of which was interpreted as an indication of the richness of the substrate coming from the root), followed by a slower phase (interpreted as a "stationary" level where substrate balances maintenance energy, and upon which migration and growth in previously uncolonised parts is superimposed). This approach allows a calculation of the basic parameter, "generation" time or "doubling" time. Comparison of generation time on roots and soil is a more correct way to define the rhizosphere effect than the R:S ratio. The generation times of *Pseudomonas* and *Bacillus* on the root were 5-6 and 39 hours respectively, and in soil they were 77 and >100 hours, respectively. The use of generation time in rhizosphere population dynamics is increasing, albeit slowly. The use of single gene mutants of bacteria is an emerging, powerful approach to questions on growth dynamics in the rhizosphere (Bowen, 1991; Lam *et al.*, 1991).

B.3.4 Interactions

In general, extrapolation from growth and interaction of organisms in laboratory media to what happens around roots is perilous (Bowen, 1980) and the studies must be performed in soil. Synergism, competition and antagonism occur in microbial growth in the rhizosphere (e.g. Bowen and Theodorou, 1979). Perception of synergism is not difficult but distinction between competition and antagonism is more difficult (but see Bowen and Theodorou, 1979).

No statement of interactions, however brief, would be complete without reference to the very neglected field of soil fauna-microflora interaction. The importance of grazing of bacteria and fungi by soil micro-fauna is becoming increasingly recognised. For example, using experimental approaches to root colonisation, Chakraborty *et al.* (1985) demonstrated that some amoebae caused 50-70% reduction of root colonisation by ectomycorrhizal fungi; Warnock *et al.* (1982) showed that grazing of vesicular arbuscular mycorrhiza hyphae by the collembola *Folsomia candida* significantly reduced the mycorrhizal plant growth response; Coleman *et al.* (1978) showed major interactions between bacteria, amoebae and nematodes affecting biomass and nutrient dynamics in soil microcosms simulating the rhizosphere.

B.3.5 The effects of the rhizosphere microflora

Non-infecting rhizosphere organisms may have several effects on plant growth. Some of these are direct and some indirect, such as the control of plant pathogens (e.g. see Rovira and Wildermuth, 1981; Wong, 1981). As indicated above, biological control of symbionts may also occur.

Unless special precautions for axenic culture are observed for control treatments, it is often extremely difficult to interpret the effects of the general rhizosphere microflora on plant growth. Various solution culture and soil culture methods have been devised for this (Barber, 1967; Rovira and Bowen, 1969). One of the greatest problems is in the use of soil, in which autoclaving sometimes produces toxins (especially in high organic matter soils). The least severe sterilising method is the use of gamma-irradiation but even this can produce some chemical changes (McLaren, 1969). Several good soil fumigation methods exist, but with these complete sterility may not occur; e.g. the use of aerated steam (Baker and Olson, 1960), methyl bromide or dazomet (for small quantities of soil) (van Berkum and Hoestra, 1979). Partial sterilisation is valuable in eliminating certain components of the soil microflora and especially where particular groups of organisms are affected, e.g. with nematicides.

The availability of nutrients such as manganese and molybdenum can be affected by the rhizosphere microflora (see Rovira *et al.*, 1983). Although phosphorus solubilisation by organisms in the rhizosphere has been proposed for some time, the ecological importance of this is still open to question, as is nutrient immobilisation. There is evidence that the rhizosphere may sometimes enhance denitrification but this is still an area for clarification (Firestone, 1982). It is also claimed that roots of some plants can inhibit nitrification in soil (Rice, 1974); this needs detailed study. Similarly, the extent of fixation of nitrogen in the rhizosphere by associative nitrogen fixers such as *Azospirillum*, and *Azobacter paspali* is still debatable due partly to inherent difficulties in field measurement. Claims for quantities fixed range from a few kg/ha/year to possibly 200 kg/ha/year. Difficulties arise from the inaccuracies of N balance methods over long periods and from extrapolation from small sample, short term incubation studies (using acetylene reduction methods). Nevertheless, the evidence indicates important accessions by associative N fixation, e.g. sometimes 30-50 kg N/ha/year in undisturbed ecosystems (e.g. grass swards) over a long period. The subject is discussed by Dobereiner (1983), Jordan (1981) and Legg and Meisinger (1982).

There are a number of indications that non-infecting rhizosphere micro-organisms produce biologically active substances affecting plants, e.g. effects on root growth and root hair production, the stimulation of intense lateral branching of some plants (called "proteoid" or "cluster" roots), the stimulation of phosphate uptake and transportation, and the induction of early flowering. The last of these can be caused by a number of organisms, of which one, *Azobacter*, was previously thought to stimulate plant growth by fixing nitrogen in the rhizosphere. Similarly, increases in plant growth sometimes caused by inoculation with *Azospirillum* may be due to hormonal effects (Beshan and Levanony, 1990). The production of plant growth promoting substances is discussed by Rovira *et al.* (1983). Such micro-organisms are referred to as Plant Growth Promoting Rhizobacteria (PGPR); responses to inoculation are often erratic. Depending on environmental factors and plant species, certain strains of rhizosphere *Pseudomonas spp.*, and some other metabolites such as HCN, may inhibit or enhance plant establishment or inhibit development of plant disease (Schippers *et al.*, 1991).

References

Asim, K.B., Steffen R.J., di Cesare, J., Haff L. and Atlas R.M. (1990) Detection of coliform bacteria in water and polymerase chain reaction and gene probes. *Applied and Environmental Microbiology* **56**, 307-314.

Baker, K.F. and Olsen, C.M. (1960) Aerated steam for soil treatment. *Phytopathology* **50**, 82.

Barber, D.A. (1967) The effects of microorganisms on the absorption of inorganic nutrients by intact plants. I. Apparatus and culture techniques. *Journal of Experimental Botany* **18**, 163-169.

Barber, D.A. and Martin, J.K. (1976) The release of organic substances by cereal roots into soil. *New Phytologist* **76**, 69-80.

Bashan, Y. and Levanony, H. (1990) Current status of *Azospirillum* inoculation technology: *Azospirillum* as a challenge for agriculture. *Canadian Journal of Microbiology* **36**, 591-608.

van Berkum, J.A. and Hoestra, H. (1979) Practical aspects of the chemical control of nematodes in soil. In: Mulder, D. (ed.), *Soil Disinfestation*, Elsevier Scientific Publishing Company, Amsterdam, pp.53-134.

Bowen, G.D. (1979) Integrated and experimental approaches to the study of growth of organisms around roots. In: Schippers, B. and Gams, W. (eds.), *Soil-Borne Plant Pathogens*, Academic Press, London, pp.209-227.

Bowen, G.D. (1980) Misconceptions, concepts and approaches in rhizosphere biology. In: Ellwood, D.C., Hedger, J.N., Latham, M.J., Lynch, J.M. and Slater, J.H. (eds.), *Contemporary Microbial Ecology*, Academic Press, London, pp. 283-304.

Bowen, G.D. (1991) Microbial dynamics in the rhizosphere: Possible stategies in managing rhizosphere populations. In: Keister, D.L. and Cregan, P.B. (eds.) *Rhizosphere and Plant Growth: Symposium Proceedings*, Kluwer Academic Publishers, Dordrecht. pp.25-32.

Bowen, G.D. and Rovira, A.D. (1973) Are modelling approaches useful in rhizosphere biology? *Bulletin of Ecological Research Communication* (Stockholm) **17**, 443-450.

Bowen, G.D. and Rovira, A.D. (1976) Microbial colonisation of plant roots. *Annual Review of Phytopathology* **14**, 121-144.

Bowen, G.D. and Theodorou, C. (1973) Growth of ectomycorrhizal fungi around seeds and roots. In: Marks, G.C. and Kozlowski, T.T. (eds.), *Ectomycorhizae - The Ecology and Physiology*, Academic Press, New York, pp.107-150.

Bowen, G.D. and Theodorou, C. (1979) Interactions between bacteria and ectomycorrhizal fungi. *Soil Biology and Biochemistry* **11**, 119-126.

Brisbane, P.G. and Kerr, A. (1983) Selective media for three biovars of Agrobacterium. *Journal of Applied Bacteriology* **54**, 425-431.

Chakraborty, S., Theodorou, C. and Bowen, G.D. (1985) The reduction of root colonization by mycorrhizal fungi by mycophagous amoebae. *Canadian Journal of Microbiology* **31**, 295-297.

Coleman, D.C., Anderson, R.V., Cole, C.V., Elliot, E.T., Woods, L. and Campion, M.K. (1978) Trophic interactions in soils as they affect energy and nutrient dynamics. IV. Flows of metabolic and biomass carbon. *Microbial Ecology* **4**, 373-380.

Dobereiner, J. (1983) Dinitrogen fixation in rhizosphere and phyllosphere associations. In: Lauchli, A. and Bieleski, R.L. (eds.), *Inorganic Plant Nutrition*. Encyclopedia of plant physiology: new ser. v. 15. Springer-Verlag, Berlin, pp.331-350.

Firestone, M.K. (1982) Biological denitrification. In: Stevenson, F.J. (ed.), *Nitrogen in Agricultural Soils*. American Society of Agronomy, Madison, Wisconsin, pp.289-326.

Foster, R.C. and Bowen, G.D. (1982) Plant surface and bacterial growth: the rhizosphere and rhizoplane. In: Mount, M.S. and Lacey, G.H. (eds.), *Phytopathogenic Prokaryotes I.* Academic Press, New York, pp.159-185

Foster, R.C. and Martin, J.K. (1981) *In situ* analysis of soil components of biological origin. In: Paul, E.A. and Ladd, J.N. (eds.), *Soil Biochemistry* **5**. Marcel Dekker, New York, pp.75-111

Foster, R.C. and Rovira, A.D. (1978) The ultrastructure of the rhizosphere of *Trifolium subterraneaum* L. In: Loutit, M.W. and Miles, J.A.R. (eds.), *Microbial Ecology*. Springer-Verlag, Berlin, pp.278-290.

Foster, R.C., Rovira, A.D. and Cock, T.W. (1983) *Ultrastructure of the Root-Soil Interface. American Phytopathology Society*, St. Paul, Minnesota.

Gilligan, C.A. (1983) Modelling of soil borne pathogens. *Annual Review of Phytopathology* **21**, 45-64.

Graham, J.H., Leonard, R.T. and Menge, J.A. (1982) Interaction of light intensity and soil temperature with phosphorus inhibition of versicular-arbuscular mycorrhiza formation. *New Phytologist* **91**, 683-690.

Harley, J.L. and Waid, J.S. (1955) A method of studying active mycelia on living roots and other surfaces in the soil. *Transactions of The British Mycology Society* **38**, 104-118.

Henry, C.M. and Deacon, J.W. (1981) Natural (non-pathogenic) death of the cortex of wheat and barley seminal roots, as evidenced by nuclear staining with acridine orange. *Plant and Soil* **60**, 255-274.

Holben, W.F. and Tiedje, J.M. (1988) Applications of nucleic acid hybridization in microbial ecology. *Ecology* **69**, 561-568.

Holden, J. (1975) Use of nuclear staining to assess rates of cell death in cortices of cereal roots. *Soil Biology and Biochemistry* **7,** 333-334.

Jordan, D.C. (1981) Nitrogen fixation by selected free-living and associative microorganisms. In: Gibson, A.H. and Newton, W.E. (eds.), *Current Perspectives in Nitrogen Fixation*. Australian Academy of Science, Canberra, pp.317-320.

Keister, D.L. and Cregan, P.B. (1991) *The Rhizosphere and Plant Growth.* Kluwer Academic Publishers, The Netherlands, 386pp.

Lam, S.F., Ellis, D.M. and Ligon, J.M. (1991) Genetic approaches for studying rhizosphere colonisation. *Plant and Soil* **129**, 11-18.

Legg, J.O. and Mersinger, J.J. (1982) Soil nitrogen budgets. In: Stevenson, F.J. (ed.), *Nitrogen in Agricultural Soils*. American Society of Agronomy, Madison, Wisconsin, pp.503-566

Lewis, D.G. and Quirk, J.P. (1967) Phosphate diffusion in soil and uptake by plants III. 31P movement and uptake by plants as indicated by 32P auto radiography. *Plant and Soil* **26**, 445-453.

Louw, H.A. and Webley, D.M. (1959) The bacteriology of the root region of the oat plant grown under controlled pot culture conditions. *Journal of Applied Bacteriology* **22**, 216-226.

McLaren, A.D. (1969) Radiation as a technique in soil biology and biochemistry. *Soil Biology and Biochemistry* **1**, 63-73.

Malajczuk, N., McComb, A.J. and Parker, C.A. (1975) An immunofluorescence technique for detecting *Phytophthora cinnamomi* Rands. *Australian Journal of Botany* **23**, 289-309.

Marschner, H. (1986) *Mineral Nutrition of Higher Plants.* Academic Press, London. 674pp.

Martin, J.K. and Kemp, J.R. (1986) The measurement of C transfers within the rhizosphere of wheat grown in field plots. *Soil Biology and Biochemistry* **18**, 103-107.

Nye, P.H. and Tinker, P.B. (1977) *Solute Movement in the Soil-Root System.* Blackwell Scientific Publications, Oxford.

Phillips, D.A., Maxwell, C.A., Hartwig, V.A., Joseph, C.M. and Wery, J. (1991) Rhizosphere flavonoids released by alfalfa. In: Keister, D.L. and Cregan, P.B. (eds.). *Rhizosphere and Plant Growth: Symposium Proceedings.* Kluwer Academic Publishers, Dordrecht, pp.149-154.

Rice, E.L. (1974) *Allelopathy.* Academic Press, London.

Riley, D. and Barber, S.A. (1969) Bicarbonate accumulation and pH changes at soyabean (Glycine max. (L.) Merr.) root-soil interface. *Soil Science Society of America Proceedings* **33**, 905-908.

Robson, A.D. (1982) Mineral nutrition. In: Broughton, W.J. (ed.), *Nitrogen Fixation* **3**. Oxford University Press, Oxford, pp.36-52.

Rovira, A.D. (1965) Interactions between plant roots and soil microorganisms. *Annual Review of Microbiology* **19**, 241-266.

Rovira, A.D. and Bowen, G.D. (1969) Phosphate incorporation by sterile and non-sterile plant roots. *Australian Journal of Biological Science* **19**, 1167-1169.

Rovira, A.D. and Davey, C.B. (1974) Biology of the rhizosphere. In: Carson, E.W. (ed.), *The Plant Root and Its Environment.* University Press of Virginia, Charlottesville, pp.153-204

Rovira, A.D. and Wildermuth, C.B. (1981) The nature and mechanisms of suppression. In: Asher, M.J.C. and Shipton, P.J. (eds.), *Biology and Control of Take-all.* Academic Press, London, pp.385-415.

Rovira, A.D., Foster, R.C. and Martin, J.K. (1979) Origin, nature and nomenclature of the organic materials in the rhizosphere. In: Harley, J.L. and Scott Russell, R. (eds.), *The Root-Soil Interface.* Academic Press, London, pp.1-4.

Rovira, A.D., Bowen, G.D. and Foster, R.C. (1983) The significance of rhizosphere microflora and mycorrhizas in plant nutrition. In: Lauchli, A. and Bieleski, R.L. (eds.), *Inorganic Plant Nutrition* (Encyclopedia of plant physiology: new ser. v. 15). Springer-Verlag, Berlin, pp.61-93.

Sarkar, A.N., Jenkins, D.A. and Wyn Jones, R.G. (1979) *Modifications to mechanical and mineralogical composition of soil within the rhizosphere* In: Harley, J.L. and Scott Russell, R. (eds.), *The Root-Soil Interface.* Academic Press, London, pp.125-136.

Schippers, B., Bakker, A.W., Bakker, P.H.A.M. and van Peer, R. (1990) Beneficial and deleterios effects of HCN-producing pseudomonads on rhizoshpere interactions. *Plant and Soil*, **129**, 75-83.

Schmidt, E.L. (1991) *Methods for microbial autecology in the soil rhizosphere.* In: Keister, D.L. and Cregan, P.B. (eds.), *Rhizosphere and Plant Growth: Symposium Proceedings*, Kluwer Academic Publishers, Dordrecht. pp.81-89.

Simon, A. and Ridge, E.H. (1974) The use of ampicillin in a simplified selective medium for the isolation of fluorescent pseudomonads. *Journal of Applied Bacteriology* **37**, 459-460.

Smiley, R.W. (1979) Wheat-rhizosphere pseudomonads as antagonists of Gaeumannomyces gaminis. *Soil Biology and Biochemistry* **11**, 371-376.

Stotzky, G. and Schenk, S. (1976) Volatile organic compounds and microorganisms. *Critical Review of Microbiology* **4**, 333-382.

Trolldenier, G. (1965) Fluoreszenmikroskipische Untersuchung von Mikroorganismenkulturen in der Rhizosphare. *Z. Bakteriol. Parasitenkd Infekt. Hyg. III* **119**, 256-259.

Tsao, P.H. (1970) Selective media for isolation of pathogenic fungi. *Annual Review of Phytopathology* **8**, 157-186.

Walker, N.A. and Smith, S.E. (1984) The quantitative study of mycorrhizal infection. II. The relation of rate of infection and speed of fungal growth to propagule density, the mean length of the infection unit and the limiting value of the fraction of the root infected. *New Phytologist* **96**, 55-69.

Warnok, A.J., Fitter, A.H. and Usher, M.B. (1982) The influence of a springtail *Folsomia candida* (Insecta, collembola) on the mycorrhizal association of the leek *Allium porrum* and the vesicular-arbuscular mycorrhizal endophyte *Glomus fasiculatus*. *New Phytologist* **90**, 285-292.

Weller, D. and Cook, R.J. (1983) Suppression of take-all of wheat by seed treatment with fluorescent pseudomonads. *Phytopathology* **73**, 463-469.

Wenzel, C.L. and McCully, M.F. (1991) Early senescence of cortical cells in the roots of cereals: How good is the evidence? *American Journal of Botany* **78**, 1528-1541.

Wong, P.T.W. (1981) Biological control by cross-protection. In: Asher, M.J.C. and Shipton, P.J. (eds.), *Biology and Control of Take-all*. Academic Press, London, pp.417-431.

Appendix C MYCORRHIZAS

David Read

C.1 INTRODUCTION

Almost all of the major plant species of natural and agroecosystems of the tropics are mycorrhizal. Literature surveys (Trappe, 1987) suggest that 85% of tropical plant species are mycorrhizal and that infection of the vesicular-arbuscular (VA) type is the most widely distributed, occurring in ca. 70% of all species. These include the major food crops of tropical regions, many of which are highly responsive to infection, at least under experimental conditions (Table C.1).

Infections of the ectomycorrhizal or sheathing type, though found on fewer species, are important because they occur on some of the most valuable timber trees of tropical forests. Orchidaceous and ericoid mycorrhizas are also found in all tropical regions, the latter mostly being restricted to high altitude situations. Approximately 13% of tropical plant species examined so far are non-mycorrhizal, these being largely confined to families all members of which are characteristically not susceptible to infection by mycorrhizal fungi such as *Caryophyllaceae, Chenopodiaceae, Commelinaceae, Cruciferae, Cyperaceae, Juncaceae, Phytolaccaceae, Polygonaceae, Portulacaceae, Proteaceae, Restionaceae and Sapotaceae.*

In the overwhelming majority of plant species that are mycorrhizal, their fungal associates occupy those distal parts of the roots which are involved in absorption of nutrients from soil. Any consideration of plant growth, crop production or ecosystem function in the tropics must therefore include an analysis of mycorrhizal biology if it is to be complete. Understanding of the processes of nutrient capture, transport and assimilation cannot be achieved without knowledge of the role of the fungal symbionts with which most plants evolved and upon which many are dependent in nature.

Assessment of the status and function of a mycorrhizal infection in any plant community normally involves a series of observations all or some of which may be required depending upon the objectives of the study. A typical series is:

1. Characterisation of infection type and analysis of occurrence of infection.
2. Quantification of occurrence.
3. Isolation and identification of fungi.
4. Inoculation of test plants and analysis of infection.
5. Analysis of host response, first in pot cultures, then in nature.

Basic procedures for such serial analysis are described below for the major mycorrhizal types.

C.2 THE VESICULAR-ARBUSCULAR MYCORRHIZAL SYMBIOSIS

This ubiquitous symbiosis is formed by zygomycetous fungi of the Order Glomales (Morton and Benny 1990). Six genera of glomalean fungi are recognized viz. *Glomus, Acaulospora, Gigaspora, Enterophospora, Sclerocystis and Scutellospora*. There are more than 150 species worldwide, most of which show low levels of host specificity. As a consequence of this it is probable that individual plants in species-rich communities such as those of savannah and humid tropical forests, most of which are heavily mycorrhizal (Newman *et al.*, 1986; Janos 1980, 1987), will be interconnected by an extensive network of VA mycelium (Read *et al.*, 1985). This mycelium, which rapidly infects the roots of seedlings, serves primarily to increase the volume of soil exploited by the plant and hence to improve its access to nutrients. The principle advantage to the plant is that ions, in particular phosphates, which diffuse slowly in soil and are readily immobilised by sorption with oxide minerals can be captured by the fungus and so rendered available (Harley and Smith, 1983).

By similar mechanisms VA fungi enhance the ability of plants to obtain phosphorus (P) released from slowly solubilizing fertilisers such as rock phosphate (Daft, 1991). Conversely, large inputs of soluble P, associated for example with application of superphosphate, can decrease or eliminate mycorrhizal advantages by inhibiting the growth and activity of the vegetative mycelium (Abbott and Robson, 1984).

C.2.1 Characterisation and analysis of VA mycorrhizal infection

The hyphae of VA mycorrhizal fungi penetrate the outer and inner cortical tissues of roots of host plants, so that 'clearing' of a sample is normally required before infection can be observed. This can be achieved by heating the sample in 10% KOH at 90°C for several minutes (Phillips and Hayman, 1970). This treatment may cause excessive disruption to delicate roots in which case an alternative method involving incubation for up to 24 hours in KOH at room temperature can be employed (Read *et al.*, 1976). In highly pigmented or suberised roots it may be necessary, following KOH treatment, to bleach tissues by immersing them in H_2O_2 (10% by volume) until they appear clear to the naked eye. After 'clearing', roots are rinsed in water, acidified with 2% HC1 and stained.

A variety of stains have been employed but, as pointed out by Grace and Stribley (1991) many of these are hazardous, some like toluidine blue: being potential carcinogens. The recommended procedure involves the use of 0.05% aniline blue or methyl blue, both names are synonyms for cotton blue in acidified 70% glycerol. The cleared root sample is immersed in the stain for 30 minutes before excess stain is removed and the sample mounted in 10% glycerine.

The diagnostic structures are the vesicles and arbuscules formed within the root cortical cells the presence of which enables the type of infection to be confirmed. The spores are formed outside the root, and enable identification of the fungi involved. Of the fungal structures, that of greatest functional significance is the arbuscule, through which nutrient exchange takes place between fungus and host. The term "arbuscular" mycorrhiza is preferable to vesicular-arbuscular because in the absence of arbuscules the mycorrhiza is unlikely to be effective.

C.2.2 Quantification of infection

The most convenient and widely used method for determining the extent of VA infection is based upon the grid intersect method (Tennant, 1975; Ambler and Young, 1977). Stained roots are laid out over a grid drawn or photocopied onto a transparent acetate sheet which can be placed on a glass slide for observation under the microscope. The numbers of fungus-grid line contacts per unit root length are determined and the result is most conveniently expressed as a proportion (usually in percentage terms) of the root length occupied by the fungus.

Table C.1 Responsiveness to VA infection of some important plänts of tropical agro and forest ecosystems.

Very responsive (i.e. Obligately mycorrhizal in most situations).	
Avocado	*Persea gratissima*
Black pepper	*Piper nigrum*
Cassava	*Manihot esculenta*
Citrus	*Citrus spp.*
Coffee	*Coffea arabica*
Cocoa	*Theobroma cacao*
Cowpea	*Vigna sinensis*
Guava	*Psidium guajava*
Mango	*Mangifera indica*
Responsive under some conditions (i.e. Facultatively mycorrhizal)	
Acacia	*Acacia spp.*
Bamboo	*Bambusa spp.*
Cotton	*Gossypium spp.*
Ground nut	*Arachis hypogea*
Litchi	*Litchi sinensis*
Maize	*Zea mais*
Mahogany	*Khaya grandifolia*
Mung bean	*Vigna radiata*
Oil palm	*Elaia guianensis*
Papaya (Paw Paw)	*Carica papaya*
Pearl millett	*Pennisetum glaucum*
Rice (dry land)	*Oryza sativa*
Rubber	*Hevea brasiliensis*
Sorghum	*Sorghum bicolor*
Sugar cane	*Saccharum officinarum*
Tea	*Camelia sinensis*
Terminalia	*Terminalia spp.*

C.2.3 Isolation and identification of VA fungi

The true status of a fungal isolate can only be assessed by back-inoculation into the host plant to confirm that structural and functional attributes of the association are reproduced (Koch, 1882). Isolation and back-inoculation is particularly important in mycorrhizal studies because fungal isolates even within a single category of mycorrhiza differ in their effects upon the nutrition and physiology of the host. Wherever possible the fungal species involved should be identified.

VA endophytes are normally isolated as spores or sporocarps which are produced in soil, especially in response to drying. The spores can be separated from the soil by wet seiving (Gerdemann and Nicolson, 1963; Pacioni, 1992) or by centifugation in 2 M sucrose (Allen *et al.*, 1979) and can be used as sources of inoculum and for purposes of identification.

In order to obtain reliable identification of spores isolated in this manner it is necessary to produce pot cultures from the single spore isolates. The spore is placed on a root of a potential host plant e.g. maize, and the two organisms are grown in a pot of previously sterilized medium which is low in phosphate for a period sufficient for infection development and spore production (3-6 months). Identification of spores can then be attempted using keys such as that of Schenck and Perez (1990). Confirmation of identity of isolates obtained in this way may be provided by INVAM: Dr Joseph Morton, Div. of Plant and Soil Sciences, 401 Brooks Hall, West Virginia University, Morganstown, WV 26506-6057, USA.

C.2.4 Inoculation of test plants

Infected roots themselves, if freshly collected from the ecosystems under investigation, can be used as sources of VA inoculum. These have the advantage that they contain the complete microbial community of the root including the naturally occurring VA endophytes. The disadvantage is that the taxonomic identity of endophytes growing in the inoculum is not known. If comparative experiments involving the growth of mycorrhizal and non-mycorrhizal plants are to be undertaken using root fragments as inoculum it is normal to add an equivalent quantity of sterilized root fragments to non-mycorrhizal treatments to balance the initial nutrient inputs. Filtrates of the root fragments containing microorganisms, but no VA fungi, may also be added to ensure that comparable non-mycorrhizal microbial communities can develop in the two treatments.

C.2.5 Analysis of host response

In all but the most fertile conditions, where depressions of host yield can occasionally be found following infection by VA fungi, a positive growth response to inoculation of sterile soil is to be expected. The range of such responses is determined not only by the fertility of the substrate but also by the extent to which the plant in question is responsive to or 'dependent' upon infection. Broadly speaking plants with fibrous root systems and well developed root hairs are less responsive than those with a coarse root system made up of tuberous roots with few root hairs. The largely non-mycorrhizal families listed above are characterised by the former type of root system. Many important plant families, for example the *Gramineae*, which also have fibrous roots, may be infected by VA fungi but be not very responsive to their presence. These infections are frequently referred to as being 'facultative'. In those plants, amongst which are numbered many tropical species of great agricultural importance, with coarse root system there may be little or no growth in nature in the absence of infection. The term 'obligate' can be applied to these infections and the host plants involved and are often said to be 'dependent' upon infection (Janos, 1987). Dependence as a concept is acceptable only if it is borne in mind that almost all autotrophs can grow independently of fungal symbionts providing their root systems are provided with an adequate supply of nutrients. The fact is that only rarely in natural ecosystems or in low external input agriculture and forestry practices are adequate supplies maintained. There are also occasions when responses to inoculation with VA fungi are greater than those obtained by application of superphosphate (Alexander *et al.*, 1992).

Various attempts have been made to quantify the extent of response of a given plant or crop to inoculation with VA fungi. Gerdemann (1975) defined responsiveness, or relative mycorrhizal dependency (RMD), as the extent to which a plant requires mycorrhizal infection to produce its maximum growth and yield at a given level of soil fertility. This concept was modified by Plenchette *et al.*, (1983) for use with crop plants under field conditions as follows. Relative field mycorrhizal dependency (RFMD) is

$$RFMD = (W_m - W_n) / W_n$$

where
W_m = Dry weight of mycorrhizal plant
W_n = Dry weight of non mycorrhizal plant

A formula for use where mycorrhizal inoculum is introduced to field soils to supplement an indigenous population (see below) has been proposed by Bagyaraj *et al.*, 1988. The mycorrhizal inoculation effect (MIE):

$$(MIE) = (W_i - W_u) / W_i$$

where
W_i = Dry weight of inoculated plant
W_u = Dry weight of uninoculated plant

This expression is useful in determining the extent to which introduced fungi compete with indigenous species to induce a growth response.

If inoculum is to be added to soil in the field, large scale production of fungal propagules is needed. This can be achieved in stages. In the first instance starter cultures are produced in pots. These can be bulked up by using progressively larger pots. The spores selected for the original inoculum may be obtained from the local site, from a culture collection or from a commercial source (e.g. Native Plants Incorp (NPI) 417 Wakara Way, Salt Lake City, UT 84108, USA or JIRA-Agroindustrias, Apartado Aereo 427, Fulua (Valle) Colombia). Concentrated inoculum already containing, for example, 1000-2000 spores of *Glomus manihotis* per ml soil can also be obtained commercially with a view to application to specific crops, in this case cassava. Such inoculum can be applied directly to sterilised seed beds on the farm to produce further bulking up. Detailed procedures for large scale inoculum production are described by Sieverding (1991), Feldmann and Idczak (1992), and Bagyaraj (1992). If pure cultures are required it is necessary to fumigate the beds prior to introduction of inoculum and seed sowing. Dazomet, applied as Basomid at 50 g/m^2 is a recommended soil fumigant. Using this, Sieverding (1991) achieved 18,314 spores of *Glomus manihotis* per gram dry soil after 6 months.

In addition to the soil based inoculum, a growth substrate consisting of porous expanded clay particles has been shown by Dehne and Backhaus (1986) to be very effective for producing vegetative mycelium and spores of VA fungi. A commercially available product Lecadan (obtainable from LECA - Germany, D 2083 Halslenbech) is recommended. Its use for inoculum production in tropical nurseries is described by Feldmann and Idczak (1992). Some clays are unsuitable because of toxic properties.

Whatever the carrier of inoculum considerable volumes are required if it is to be applied on a field scale. Sieverding (1991) used 5000 litres of inoculum, consisting of roots, mycelium and spores in soil, per ha of cassava. Even with this level of application, using a highly responsive plant, and a fungus selected for its effectiveness, benefits may not persist into the second rotation. Loss of benefit can be attributable to failure of the introduced inoculum to compete with indigenous endophytes or simply to the fact that the latter are better adapted and so persist longer. With the exception of severely eroded or disturbed sites, and those that have been left as open fallow for a number of years most soils will contain a native inoculum with attributes that are likely to be well suited to the site (see Dodd *et al.*, 1983). This being the case the first question to be asked in any situation involving the possibility of inoculation is not so much which inoculum to select but whether inoculation is indeed the appropriate option (see Abbott *et al.*, 1992). The complexity of the inoculum production process and the uncertainties of success are such that it may be preferable to consider manipulating the native inoculum to maximise the benefits obtained from it.

In crop rotation systems, and in intercropping, the use of non-mycorrhizal crop species such as chenopods or crucifers (Harinikumar and Bagyaraj, 1988) will lead to a reduction in numbers of infective progagules in soil. In contrast, use of a crop such as cassava which is strongly mycorrhizal will increase their number, as will the use of a grass rather than a non-mycorrhizal weed in any fallow system. Comparisons of row intercropping with sole cropping of maize and bean have shown higher rates of infection and yields of both plants in the intercropping system (Sieverding, 1991).

Pre-sowing cultural practices can also be modified to improve inoculum vigour. Addition of organic manure (Harinikumar and Bagyaraj, 1989) or straw (Daft, 1992) can increase the size and effect of VA inoculum in cropped systems, whereas heavy doses of super-phosphate and tillage will both reduce it. By fragmenting the mycelial network established under the previous crop, tillage can be expected to reduce its inoculum potential. Evidence from a temperate agricultural cropping system involving maize now points clearly to such effects. Mycorrhizal infection, phosphorus inflow and growth of the crop were all shown to be significantly greater in a no-till than in a ploughed system (McGonigle *et al.*, 1990; Miller and McGonigle, 1992). Within slash and burn agriculture there appears to be little adverse effect on propagule density of VA fungi (Sieverding, 1991).

C.3 THE ECTOMYCORRHIZAL SYMBIOSIS

Most ectomycorrhizal fungi are basidiomycetes but there are some ascomycetous associations. This symbiosis is found in the majority of woody species of the temperate and boreal forests and in some of the most important tropical tree species. Members of the Dipterocarpaceae, which make up ca. 80% of timber exported from South east Asia and some 30% of the total tropical hardwood trade (Smits 1992) appear to be almost entirely ectomycorrhizal, as are the Fagaceae, the sub-family Leptospermoideae of the Myrtaceae, which includes *Eucalyptus*, and some genera of the tribes Amhersteae and Detariae in the legume sub-family Caesalpinoideae (Alexander, 1989a). Non-nodulated ectomycorrhizal trees conspicuously dominate in many moist savannas including the genera *Brachystegia, Isoberlinia, Julbernardia, Marquesia, Monotes and Uapaca* in Africa (Alexander and Hogberg, 1986; Hogberg, 1989) Selection favouring ectomycorrhizal species is thought to be driven by nutritional factors which ultimately determine community composition. Both in Amazonian (Singer and Araujo, 1979; 1986) and West African (Newbery *et al.*, 1988) forests, communities dominated by ectomycorrhizal species are characteristically restricted to extremely nutrient-poor soils with surface accumulation of litter. There is increasing evidence that this symbiosis is involved in the mobilisation of nitrogen as well as phosphorus from these substrates (Read, 1991).

In contrast to the VA fungi, the ectomycorrhizal fungi in tropical forests, show a greater level of host-specificity (Thoen and Ba, 1987; Alexander, 1989b; Smit, 1992). This has implications for diversity of their host species. Alexander (1989b) points out that in the more host specific circumstance the transfer of infection from established mycorrhizal mycelium to germinating seedling while still being effective, would lead to a diametrically opposed effect to that seen in VA forests of low host specificity. By increasing the survivorship of a small number of compatible species, diversity is actually reduced. There is some evidence from both the Amazonian and African forests that species diversity is lower in ecto than in adjacent VA forests.

C.3.1 Characterisation and analysis of ectomycorrhizal infection

This mycorrhizal type is characterised by inter-cellular as distinct from intra-cellular penetration by the mycelium to form a 'Hartig-net' through which nutrient exchange between fungus and host occurs. There is also a more or less well developed 'sheath' or 'mantle' of fungal mycelium around the root which may have a pseudoparenchymatous or plectenchymatous structure.

Ectomycorrhizal lateral roots are often short and swollen and so can be identified with the naked eye. Confirmation of infection requires that a longitudinal or transverse section be taken to reveal the Hartig-net and the ensheathing layer or mantle of mycelium, which can be of variable thickness around the whole root.

C.3.2 Quantification of occurrence

Methods for sampling ectomycorrhizal fine roots and of quantifying their occurrence are described by Fairley and Alexander (1985).

C.3.3 Isolation and identification of ectomycorrhizal fungi

Sporocarps provide identifiable sources of ectomycorrhizal fungi, and many species are readily grown from vegetative parts of young sporocarps. Spores provide less reliable sources of inoculum, often requiring roots or root exudates to stimulate germination. Isolates can also be obtained by surface sterilisation of root tips using 30% H_2O_2 for 2 minutes (Taylor and Alexander, 1991), but it is difficult to ascribe such isolates to particular species. Detailed analyses enabling tentative identification of the causal fungus require the study of root squashes or scrapings of the fungal mantle which can reveal diagnostic features of mycelial structure (Agerer, 1987; 1990; 1991; Ingleby *et al.*, 1990).

C.3.4 Inoculation of test plants

Detailed methods for synthesis of ectomycorrhiza under laboratory conditions are provided by Peterson and Chakravarty (1991).

Field inoculation has often been attempted and particularly striking responses have been achieved in the tropics when alien species such as pine and eucalyptus have been introduced into areas which lack appropriate ectomycorrhizal fungi. One fungus, above all others, *Pisolithus tinctorius*, has proved to be useful as an inoculant for use on seedlings to be transplanted from nurseries into tropical soils. The success of this fungus appears to be associated with a combined ability to tolerate high temperature and to produce a very large and vigorous external mycelium. A country by country account of the successes achieved by inoculation with this fungus together with details of production and inoculation methods is provided by Marx (1991). Inoculum of *P. tinctorius* and other ectomycorrhizal fungi is commercially available from Mycorr Tech. Inc. University of Pittsburgh Applied Research Centre, Pittsburgh, Pennsylvania 15238, USA. *P. tinctorius* shows relatively little host specificity.

References

Abbott, L.K., Robson, A.D. and DeBoer, G. (1984) The effect of phosphorus on the formation of hyphae in soil by the vesicular-arbuscular mycorrhizal fungus *Glomus fasciculatum*. *New Phytologist* **97**, 437-446.

Abbott, L.K., Robson, A.D. and Gazey, C. (1992) Selection of inoculant vesicular-arbuscular mycorrhizal fungi. In: Norris, J.R., Read, D.J. and Varma, A.K. (eds.), *Methods in Microbiology* **24**. Academic Press, London, pp.1-21.

Agerer, R. (ed.). (1987-1990) *Colour atlas of ectomycorrhizae* 1st-4th Edition. Einhorn-Verlag, Schwabisch Gmund, Germany.

Agerer, R. (1991) Characterisation of ectomycorrhiza. In: Norris, J.R., Read, D.J. and Varma, A.K. (eds.), *Methods in Microbiology* **23**. Academic Press, London, pp.25-73.

Alexander, I.J. (1989a) Systematics and ecology of ectomycorrhizal legumes. In: Stirton, C.H. and Zarucchi, J.L. (eds.), *Advances in legume biology*. Monograph of Missouri Botanical Gardens, pp.607-624.

Alexander, I.J (1989b) Mycorrhizas in tropical forest. In: Proctor, J. (ed.), *Mineral Nutrients in Tropical Forest and Savanna Ecosystems*. Blackwell Scientific Publishers, Oxford, pp.169-188.

Alexander, I.J. and Hogberg, P. (1986) Ectomycorrhizas in tropical angiosperm trees. *New Phytologist* **102**, 541-549.

Alexander, I.J., Ahmad, N. and See, L.S. (1992) The role of mycorrhizas in the regeneration of some Malaysian forest trees. *Philosophical Transactions of the Royal Society* **335** (in press).

Allen, M.F. (1991) *The ecology of mycorrhizae.* Cambridge University Press, London.

Allen, M.F., Moore, T.S. and Christensen, M. (1979) Growth of vesicular-arbuscular mycorrhizal and non mycorrhizal *Bouteloua gracilis* in a defined medium. *Mycologia* **71**, 666-669.

Ambler, J.R and Young, J.L. (1977) Techniques for determining root length infected by vesicular-arbuscular mycorrhizae. *Soil Science Society of America Journal* **41**, 551.

Bagyaraj, D.J. (1992) Vesicular-arbuscular mycorrhiza : Application in Agriculture. In: Norris, J.R., Read, D.J. and Varma, A.K. (eds.), *Methods in Microbiology* **24**. Academic Press, London, pp. 359-373.

Daft, M.J. (1991) Influences of genotypes, rock phosphate and plant densities on mycorrhizal development and the growth of five different crops. *Agriculture, Ecosystems and Environment* **35**, 151-169.

Daft, M.J. (1992) Use of VA mycorrhizas in agriculture: Problems and Prospects. In: Read, D.J., Lewis, D.H., Fitter, A.H. and Alexander, I.A. (eds.), *Mycorrhizas and Ecosystems*. CAB International, Wallingford, UK. pp.198-201.

Dehne, H-W and Backhaus, G.F. (1986) The use of vesicular-arbuscular mycorrhizal fungi in plant production. I. Inoculum production. *Journal of Plant Diseases and Protection* **93**, 415-424.

Dodd, J.C., Krikun, J. and Hass, J. (1983) Relative effectiveness of indigenous populations of vesicular-arbuscular mycorrhizal fungi from four sites in the Negev, Israel. *Journal of Botany* **32**, 10-21

Fairley, R.I. and Alexander, I.J. (1985) Methods of calculating fine root production in forests. In: Fitter, A.H., Atkinson, D., Read, D.J. and Usher, M.B. (eds.), *Ecological Interactions in Soil, Plants, Microbes and Animals.* British Ecological Society Special Publication 4. Blackwell, Oxford, pp.37-42.

Feldmann, F. and Idczak, E. (1992) Inoculum production of vesicular-arbuscular mycorrhizal fungi for use in tropical nurseries. In: Norris, J.R., Read, D.J. and Varma, A.K. (eds.), *Methods in Microbiology* **24.** Academic Press, London, pp.339-357.

Gerdemann, J.W. (1975) Vesicular-arbuscular mycorrhizae. In: Torrey, J.G. and Clarkson, D.T. (eds.), *The Development and Function of Roots.* Academic Press, New York, pp.575-591.

Gerdemann, J.W. and Nicolson, T.H. (1963) Spores of mycorrhizal *Endogone* species extracted from soil by wet sieving and decanting. *Transactions of the British Mycological Society* **46**, 235-244.

Grace, C. and Stribley, D.P. (1991) A safer procedure for routine staining of vesicular-arbuscular mycorrhizal fungi. *Mycological Research* **95**, 1160-1162.

Harinikumar, K.M. and Bagyaraj, D.J. (1988) The effect of crop rotation on native vesicular-arbuscular mycorrhizal propagules in soil. *Plant and Soil* **110**, 77-80.

Harinikumar, K.M. and Bagyaraj, D.J. (1989) Effects of cropping sequence, fertilisers and farmyard manure on vesicular-arbuscular mycorrhizal fungi in different crops over three consecutive seasons. *Biology and Fertility of Soils* **7**, 173-175.

Harley, J.L. and Smith, S.E. (1983) *Mycorrhizal symbiosis.* Academic Press, London.

Hogberg, P. (1989) Root symbioses of trees in savannas. In: Proctor, J. (ed.), *Mineral Nutrients in Tropical Forest and Savanna Ecosystems.* Blackwell Scientific Publications, Oxford, pp.121-137.

Ingleby, K., Mason, P.A., Last, F.T. and Fleming, L.V. (1990) *Identification of Ectomycorrhizas.* H.M.S.O. London.

Janos, D.P. (1980) Vesicular-arbuscular mycorrhizas affect lowland tropical rain forest plant growth. *Ecology* **61**, 151-162.

Janos, D.P. (1987) VA Mycorrhizas in humid tropical ecosystems. In: Safir, G.S. (ed.), *Ecophysiology of VA Mycorrhizal Plants.* CRC Press, Boca Raton, LA, USA, pp.107-135.

Koch, R. (1882) Uber die Milzbrandimpfung, eine Entgegnung auf den von Pasteur in Genf gehaltenen Vortrag. Fischer, Berlin.

Marx, D.H., Ruehle, J.L. and Cordell, C.E. (1991) Methods for studying nursery and field response of trees to specific ectomycorrhiza. In: Norris, J.R., Read, D.J. and Varma, A.K. (eds.), *Methods in Microbiology* **23**. Academic Press, London, pp.383-411.

McGonigle, T.P., Evans, D.G. and Miller, M.H. (1990) Effect of degree of soil disturbance on mycorrhizal colonisation and phosphorus absorption by maize in growth chamber and field experiments. *New Phytologist* **116**, 629-636.

Miller, M.H. and McGonigle, T.P. (1992) Soil disturbance and the effectiveness of arbuscular mycorrhiza in an agricultural ecosystem. In: Read, D.J., Lewis, D.H., Fitter, A.H. and Alexander, I.J. (eds.), *Mycorrhizas in Ecosystems.* CAB International, Wallingford, UK. pp.156-163.

Morton, J.B. and Benny, G.L. (1990) Revised classification of arbuscular mycorrhizal fungi (Zygomycetes): a new order, Glomales, two new suborders, Glomineae and Gigasporineae and two new families Acaulosporaceae and Gigasporaceae with an emendation of Glomaceae. *Mycotaxon* **37**, 471-491.

Newbery, D.M., Alexander I.J., Thomas, D.W, and Gartlan, J.S. (1988) Ectomycorrhizal rain-forest legumes and soil phosphorus in Korup National park, Cameroon. *New Phytologist* **109**, 433-350.

Newman, E.I., Child, R.D. and Patrick, C.M. (1986) Mycorrhizal infection in grasses of Kenyan savanna. *Journal of Ecology* **74**, 1179-1183.

Pacioni, G. (1992) Wet sieving and decanting techniques for the extraction of spores of vesicular-arbuscular fungi. In: Norris, J.R., Read, D.J. and Varma, A.K. (eds.), *Methods in Microbiology* **24**. Academic Press, London, pp.317-323.

Peterson, R.L. and Chakravarty, P. (1991) Techniques in synthesizing ectomycorrhizas. In: Norris, J.R., Read, D.J. and Varma, A.K. (eds.), *Methods in Microbiology* **23**. Academic Press, London, pp.75-106.

Phillips, J.M. and Hayman, D.S. (1970) Improved procedures for cleaning roots and staining parasitic and vesicular-arbuscular mycorrhizal fungi. *Transactions of the British Mycological Society* **55**, 158-160.

Plenchette, C., Fortin, J.A. and Furlan, V. (1983) Growth response of several plant species to mycorrhiza in a soil of moderate P- fertility. I. Mycorrhizal dependency under field conditions. *Plant and Soil* **70**, 199-209.

Read, D.J. (1991) Mycorrhizas in ecosystems. *Experientia* **47**, 379-391.

Read, D.J., Koucheki, H.K. and Hodgson, J. (1976) Vesicular-arbuscular mycorrhiza in natural vegetation systems. I. The occurrence of infection. *New Phytologist* **77**, 641-653.

Read, D.J., Franics, R. and Finlay R.D. (1985) Mycorrhizal mycelia and nutrient cycling in plant communities. In: Fitter, A.H., Atkinson, D., Read, D.J. and Usher, M.B. (eds.), *Ecological Interactions in Soil: Plants, Microbes and Animals.* British Ecological Society, Special Publication **4**. Blackwell, Oxford, pp.193-217.

Schenck, N.C. and Perez, V. (1990) *Manual for the Identification of VA Mycorrhizal Fungi.* Synergistic Publications, Gainesville, Florida, USA.

Sieverding, E. (1991) Vesicular-arbuscular mycorrhiza management in tropical agroecosystems. Hartmit Bremer Verlag. 3403 Friedland 5, Germany, 371 pp.

Singer, R and Araujo, I. (1979) Litter decomposition and ectomycorrhiza in an Amazonian forest. I. A comparison of litter decomposing and ectomycorrhizal basidiomycetes in latosol-terra-firme forest and white podzol campinarana. *Acta Amazonica* **9**, 25-41.

Singer, R. and Araujo, I. (1986) Litter decomposition and ectomycorrhizal basidiomycetes in an igapo forest. *Plant Systematics and Evolution* **153**, 107-117.

Smits, W. (1992) Mycorrhizal studies in Dipterocarp forests in Indonesia. In: Read, D.J., Lewis, D.H., Fitter, A.H. and Alexander, I.J. (eds.), *Mycorrhizas in Ecosystems.* CAB International, Wallingford, UK. pp.283-292.

Taylor, A.F.S. and Alexander, I.J. (1991) Ectomycorrhizal synthesis with *Tylospora fibrillosa* a member of the Corticiaceae. *Mycological Research* **95**, 381-384.

Tennant, D. (1975) A test of a modified line intersect method of measuring root length. *Journal of Ecology* **63**, 995-1001.

Thoen, D. and Ba, A.M. (1987) Observations on the fungi and the ectomycorrhiza of *Afzelia africana* and *Uapaca guineensis* in Southern Senegal. In: Sylvia, D.M., Hung, L.L., Graham, J.H. (eds.), *Proceedings of the 7th North American Conference on Mycorrhiza.* University of Florida Press, Gainesville, Florida, USA, p.132.

Trappe, J.M. (1987) Phylogenetic and ecological aspects of mycotrophy in the angiosperms from an evolutionary standpoint. In: Safir, G.R. (ed.), *Ecophysiology of VA Mycorrhizal plants.* CRC Press, Boca Raton, LA, USA, pp.5-25.

Appendix D ROOTS: LENGTH, BIOMASS, PRODUCTION AND MORTALITY

Meine van Noordwijk

D.1 INTRODUCTION

In both crops and natural vegetation a substantial proportion of net primary production occurs below ground in the root system. This proportion varies with vegetation type, developmental stage, soil conditions and cultural practices. Shoot:root ratios on a dry matter basis are typically between 5:1 and 10:1 for annual crops at maximum standing biomass. For perennial crops and natural vegetation, values vary between 1:2 and 5:1. Uptake of water and nutrients are related to root length or root surface area rather than root weight. Distribution and periodicity of root length is important to evaluate whether or not crop demand for nutrients and water coincides in space and time with the available supply (synchrony and synlocalization). Root death and root exudates are a major input of organic matter to the soil system and the extent, timing and location of root death are therefore important. By leaving a well distributed set of continuous channels of mostly easily decomposable organic matter, roots have a relevance for soil biota, including roots of subsequent crops, far beyond their often limited quantity of organic matter.

For these reasons root investigations are an important part of the TSBF programme. Unfortunately they present particular difficulties; extraction of roots from soil is time consuming, labour intensive and still it is often incomplete.

D.2 QUANTIFICATION OF ROOT BIOMASS AND LENGTH

The methods described in Section 3.3 of this Handbook may, with adequate calibration, be used for estimates of root weight and length. More accurate estimates can be obtained from well replicated core (auger) samples, washed on a fine meshed sieve. To obtain reliable estimates of root biomass several points should be noted:

- samples should be taken from representative volumes of soil; in row crops special sampling schemes (Van Noordwijk *et al.*, 1985) may be needed;
- samples should be taken around the expected maximum standing root biomass; a late season sampling may result in a high proportion of dead roots;
- the methods for sampling, storing samples and washing will unavoidably lead to some loss of dry weight and nutrients; relevant correction factors may be obtained by simulating all procedures on roots grown in nutrient solution. Losses can be in the range 20 - 50% for dry weight, and thus correction factors of 1.25 - 2.0 should be applied to the final data (Van Noordwijk, 1987; Grzebisz *et al.*, 1989). Nutrient contents are generally only reliable if sample handling is completed within one day.

D.2.1 Corer design

The corer removes a known volume of soil from a known depth in the profile, without the need for digging a soil pit and destroying part of an experimental field. A core of 50 - 80 mm diameter is satisfactory, and the corer can be inserted either manually or mechanically. Manual coring is difficult at depths greater than 50 cm and in clay or stony soil. In dry sandy soil a smaller core diameter may be needed to reduce losses of soil when extracting the core. A suitable hand corer consists of a 15 cm steel tube with a serrated cutting edge mounted on a 1 m pipe, with a plunger to remove the core (Figure D.1; Bohm, 1979). Marks on the pipe indicate 10 cm depth increments. Soil is extracted with the corer in successive 10 cm increments.

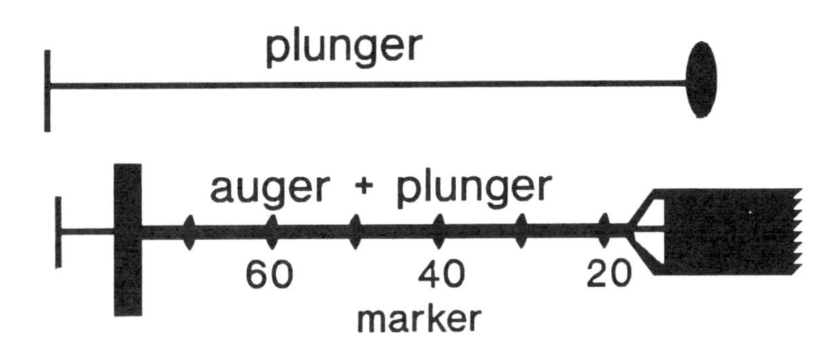

Figure D.1 Design of a hand corer

Alternatively a sharpened steel tube of appropriate length can be driven into the soil. In this case it is important to have a tripod and chain hoist to extract the tube, a file to resharpen the cutting edge, and longitudinally split liners of thin metal within the tube so that the intact soil core can be pulled out. A powered motor breaker can be used to drive the corer into the soil (Welbank and Williams, 1968), but compaction of the core will occur and should be allowed for when cutting subsamples from the core.

In very stony soil, or where there are many woody tree roots, coring may not be possible. Regular, known volumes of soil (monoliths) can be taken from the face of a pit and treated in the same way as cores.

D.2.2 Sample depth

Ideally the profile should be sampled to the limits of rooting depth. At depth, however, rooting intensity is low and spatial variability high. Based on initial profile wall observations a meaningful lower limit can be set. In some cases a linear relationship of the log of root mass versus depth (a negative exponential root distribution) may help to extrapolate root densities in the soil beyond sampling depth. All soils must be sampled to a minimum depth of 30 cm.

Where horizon development occurs, subdivide the cores at horizon boundaries of known depth in the first instance and within horizons in 10 cm increments. In cultivated soil or where there are no clear horizons subdivide the mineral soil in 10 cm increments. Surface organic horizons should be treated separately.

D.2.3 Sampling intensity

Even in the most homogeneous soils a considerable spatial variability of root density will occur. For auger samples of about 385 cm^3 (10 cm height, 7 cm diameter) a coefficient of variation in root weight of at least 40% may be expected (Van Noordwijk *et al.*, 1985a); on heterogeneous soils the coefficient of variation may be much higher. This variability implies that large number of replicate samples are needed if precise estimates of root weight are needed.

It is advisable to obtain reliable information at one or two, well chosen situations, rather than non-reliable data on many. If 25 replicates would be analysed the standard error of the mean would be five times smaller than the standard deviation of individual samples, and thus the 95% confidence interval of the mean would be plus or minus 20%, even in the most homogeneous soil. Within each replicate plot of a treatment take no fewer than 3 cores. Within each replicate the samples at each depth increment can be pooled for further treatment. In natural vegetation where no sample stratification strategy is obvious, take the cores on random coordinates. Where patterns are likely to occur (e.g. row crops, alley cropping) the first stratification should be on an area basis (within rows vs. between rows), then sample randomly within the strata. Beware of high root mass directly underneath the plants. Root data are seldom normally distributed and an appropriate transformation ($(n+1)^{0.5}$ or log n) is often required before assessing the significance of differences between treatments or sampling intervals.

D.2.4 Root extraction

The best approach is to wash roots from the cores immediately upon return from the field. Core samples can be stored in sealed polythylene bags in a refrigerator for a few days or deep freeze until processed. If deep freeze facilities are not available, samples can be stored air-dried and re-wetted before washing. Losses of dry weight due to the methods used for storage should be checked.

Soil texture, structure, degree of compaction and organic matter content greatly influence the precision and time required to extract roots from cores. The simplest method involves gently washing a presoaked sample over a large diameter 0.3 - 0.5 mm mesh sieve. The work can be simplified by washing over a combination of sieves: one with 1.1 and one with 0.3 mm mesh. The first sieve will contain mostly roots, the second mostly debris. The material removed from the sieve(s) can then be mixed in water and the suspended material decanted (live roots have a specific density of about 1.0 g/cm^3). This residue is then hand sorted in shallow dishes under water to remove fragments of organic matter and dead roots; normally it is better to pick live roots from the sample and leave debris behind in the dish. Good light conditions, a calliper with 0.1 mm accuracy and a pair of (watchmaker) forceps are necessary; a stereo dissecting microscope (from x 4 to x 20) may be helpful. Operator fatigue is a problem and it may take about 6 hours to sort one core (to 1 m depth) from maize on a sandy loam, a similar core from forest with a high organic matter content or just one sample of the top layer of permanent grassland.

A number of root washing machines have been designed (Bohm, 1979). The most successful employ the process of elutriation, i.e. washing roots and organic debris free of soil and separating them by flotation, onto a 0.5 mm mesh sieve leaving behind the heavier mineral particles. The apparatus of Cahoon and Morton (1981) requires adequate water pressure and uses a lot of water; that of Smucker *et al.* (1982), now commercially available, uses less water and accepts smaller samples but requires both water pressure and compressed air. After elutriation roots must still be sorted by hand and this may take several hours. Neither apparatus handles organic soils. With both hand and machine washing, loss of fine roots occurs and a periodic check of washing water and residues should be made to quantify such losses; such losses should be kept less than 5% of root weight (which may imply that they are still over 10% of root length); this loss is distinct from that due to respiration and loss of cell contents from remaining root tissue.

Presoaking overnight in 5% sodium hexametaphosphate expedites the process of washing roots from clay soils, but the chemical discolours the roots, particularly in soils with high organic matter content and may disrupt the tissue, making subsequent identification of live roots more difficult. Such pretreatment will also interfere with chemical analyses and should therefore be avoided if possible. Any lengthy washing procedure may alter the element content of root tissue and a subsample hand sorted with a minimum of water and processed on the day of sampling must be used for analysis.

D.2.5 Classifying the roots

Fine roots are the most important part of the root system for water and nutrient uptake, as they form the largest part of total root length or root surface area. For woody perennial vegetation there is a fairly obvious distinction between the more or less permanent, secondarily thickened roots and the ephemeral, unthickened roots. This functional distinction usually falls somewhere between 1 and 3 mm root diameter. Roots above 10 mm diameter are not adequately sampled by coring. For herbaceous perennial and short lived vegetation no such clear distinction exists. For TSBF studies these roots should be separated into <2 mm and >2 mm classes. In mixed vegetation, separation of roots of different species may be difficult and is not necessary for TSBF research.

It is desirable to separate root samples into living and dead categories. This is particularly important in crop situations where old, dead roots from the previous crops may still be present. Living roots can be distinguished by their lighter colour, turgid appearance and flexible rather than friable nature when manipulated. Some preliminary anatomical investigation may help in establishing criteria for making decisions. Incubation of excised roots in soil in modified litter bags can be used to establish visual clues to the root decay process. Cross checking between operators and working block by block instead of treatment by treatment help to reduce experimental error.

Congo red staining has been used to differentiate wheat roots with intact epidermis from ones without (Ward *et al.*, 1978). Stain in 1% aqueous solution of Congo Red for 3 min, rinse, blot dry, then saturate for 3 min in 98% ethanol before a final rinse. Living roots stain dark pink to bright red. The criteria for making the living/dead distinction must be clearly stated. Where adequate criteria cannot be developed, assess total root mass only.

D.2.6 Assessment of root mass

Washed root samples can be stored in sealed polyethylene bags for a short time in a refrigerator, but deep-freeze storage is preferable. Thymol can be added as a bactericide, but should be handled carefully; classical storage media (ethanol, etc.) tend to make roots brittle.

Carry out biomass estimation on each size class of all samples. Oven-dry the roots and weigh. Next the dried samples should be combusted for 5 hr in a muffle furnace at 550°C and the residue weighed. Results should be expressed as ash-free oven-dry mass per unit volume of soil.

D.2.7 Root length measurements

Root length is a relevant parameter for water and nutrient uptake. The specific root length (length per unit dry weight) of roots depends on diameter, variability of diameters, dry matter content (per unit fresh weight) and air-filled porosity (per unit volume) (van Noordwijk, 1987). Within a species or situation variability of the specific root length may be rather small (e.g. with a coefficient of variation of 10-15% while root weight per unit soil volume has a coefficient of variation of about 40%), so it is reasonable to measure the specific root length only for some subsamples. Normal values are from about 10 m/g for fine tree roots, 50-200 m/g for fine roots of dicotyledonous crops and 200-600 m/g for cereals and grasses (Van Noordwijk and Brouwer, 1991).

Root length can be estimated by counting the number of intersections between roots and sample lines. This method is based on Buffon's needle problem, described in 1777, where the chance that a needle randomly thrown on a tiled floor would intersect one or more of the edges of the tiles was formulated as a function of the length of the needle and the size of the tile. Application of the method for measuring root length is based on Newman (1966) and Tennant (1975). Roots are spread out with random orientation in a thin layer of water on a glass plate (about 25 x 25 cm), water is removed and a grid (photocopied on an acetate folio for overhead projection) is put over or underneath the sample. Line by line all intersections of roots and grid lines (taking the upper or left boundary of the line as criterion in case of doubt) are added (Figure D.2). Results for horizontal (H) and vertical (V) grid lines are added to the number N. If the grid size is D (mm), root length L (mm) is derived as:

$$L = \pi N D / 4$$

If D is set equal to $40/\pi$, i.e. 12.7 mm, L = 10 N (mm). By adding results for H and V lines the method is insensitive to preferential oreintation of roots on the plate, but spreading the roots must be done without regard of the position of the grid lines. To improve contrast roots can be stained beforehand, e.g. with saffranin red (1 g/litre).

To optimise working efficiency at an acceptable random error level the grid size should be chosen to obtain about 400 intercepts (200 H + 200 V) per sample. This method can be applied manually (using a magnifying lens for fine roots) or can be automated in various ways (Rowse and Phillips, 1974; Richards *et al.*, 1979; Wilhelm *et al.*, 1982). Each variant of the method should be calibrated by cutting a known length of cloth to small pieces. Commercial equipment is available (Comair Root Length Scanner, Commonwealth Aircraft Corporation

Ltd, 304 Lorimer St, Port Melbourne, Victoria 32307, Australia). Recently computer image analysis methods are used as well, based on a video camera or line scanner. Only small fields of view can be analysed with sufficient resolution and root samples should be spread out more carefully than when human eyes are used as image analyser, so the gain may be less than expected. For TSBF purposes manual versions of the method are recommended.

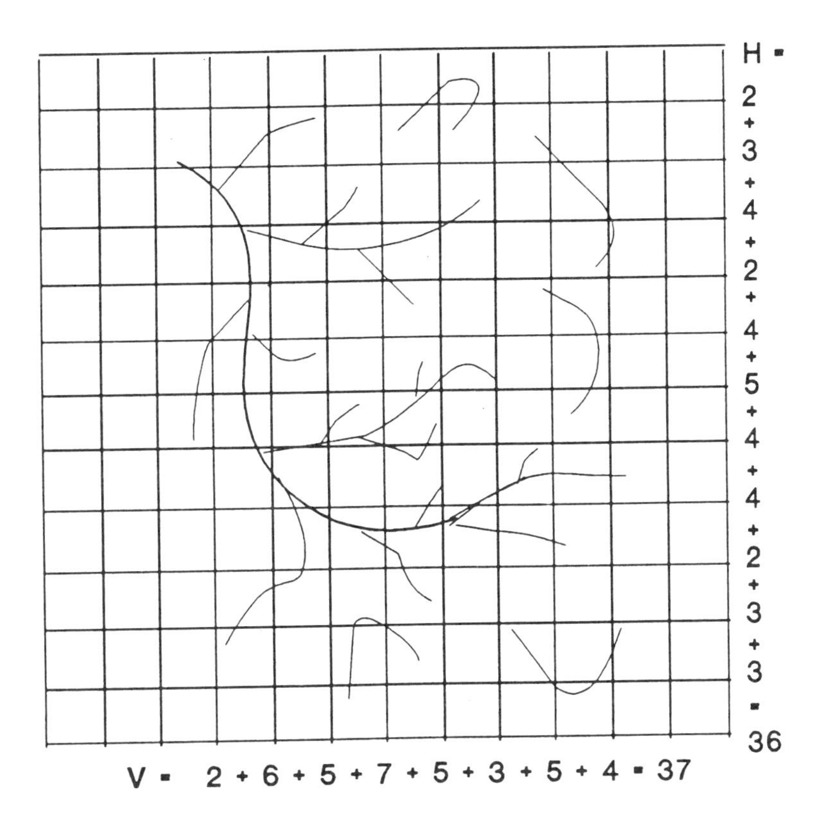

Figure D.2 Line intercept method for determining root length by counting the number of interceptions between roots and horizontal (H) plus vertical (V) lines of a grid.

Measurements of the frequency distribution of root diameters can be made on the sample spread out on a grid, by measuring on every *x*th interception, using a binocular microscope with an ocular micrometer. At least 20 readings per sample are required (assuming three samples per treatment per layer).

D.2.8 Mycorrhizas

It is desirable to have some measure of the type and intensity of mycorrhizal infection in each treatment. At peak root biomass take approximately 60 cm of fresh clean root at random from each depth and stain to visualize fungal hyphae. The method of Phillips and Hayman (1970) can be followed with small adaptations: short (ca. 2 cm) lengths of roots (stored in alcohol-acetic acid mixture after washing) are heated for about 1 hr at 90°C in 10% KOH (or left for 2-3 days at ambient temperature), washed in water, if necessary (dark tissue) bleached in alkaline peroxide (3% NH_4OH in 3% H_2O_2, prepare daily), washed in water, acidified for 3 min in 1% HCl, stained for 5 min at ambient tempearture in 0.05% trypan blue (Merck Art No. 11732) in lactic acid and destained in clear lactic acid. In the original description lactophenol (1:1:1:1 mixture of phenol, lactic acid, glycerol and water) was used instead of lactic acid. As phenol is toxic it should not be used, unless staining results are unsatisfactory. A modified procedure was described by Kormanik and McCraw (1982). For roots with a high lignin content the KOH treatment may have to be intensified. After staining the samples are inspected under a (dissecting) microscope. Vesicular arbuscular mycorrhiza (VAM) infection is characterized by the formation of unseptated hyphae outside the root and inter- and/or intracellular hyphae in the cortical cell layers of the root (Sieverding, 1991). The percentage of root sections which has mycorrhizal structures is assessed. Ectomycorrhizas are characterized by a fungal sheath and/or Hartig net; the percentage of root tips with such structures (which can often be recognized macroscopically and without staining) is assessed.

See also Appendix C "Mycorrhizas".

D.2.9 Sampling frequency

The amount of living and dead root in the ecosystem fluctuates as a balance of new root growth and root death and decay. Given the large number of replicate samples required to obtain a reasonable (let alone accurate) estimate of root density at a given moment, a regular sampling programme to monitor changes during the growing season easily becomes unmanageable. For TSBF site characterization a minimum of two samples per year is required.

For perennial vegetation some prior knowledge of the phenology of the root system will allow sampling to coincide with the likely peaks and troughs in root biomass ; root growth often alternates with periods in which reproductive growth is a major sink for carbohydrates. Where no prior knowledge exists samples should coincide in herbaceous vegetation with maximum and minimum above-ground biomass, and in forest with the months of minimum and maximum rainfall. For annual crops maximum root development often coincides with flowering (transition from vegetative to generative stage).

Sampling at harvest time may lead to a considerable amount of dead roots. Sampling before planting a new crop is relevant for estimating the rate of decay of dead roots from previous crops.

D.3 ESTIMATION OF TOTAL ROOT PRODUCTION

To estimate total root production the biomass estimates obtained (corrected' for sampling schemes and dry matter losses) should be corrected for root turnover between sampling dates. Due to the large variability of results from destructive sampling it is not possible, in practice, to obtain turnover estimates from frequent sampling schemes.

The frequent sampling method derives from the following scheme with two pools (live and dead roots) and three rates (production, mortality and disappearance):

Production (R_p) --> Live root mass (L_t) --> Mortality (R_m) --> Dead root mass (D_t) --> Disappearance (R_d)

The following equations hold:

$$L_{t+1} = L_t + \int R_p \; - \int R_m$$

$$D_{t+1} = D_t + \int R_m \; - \int R_d$$

If live and dead roots cannot be separated:

$$(L+D)_{t+1} = (L+D)_t + \int R_p \; - \int R_d$$

Thus:

$$\int R_p \geq (L+D)_{t+1} - (L+D)_t$$

Sampling errors in determining L_t and D_t play a dominant role, however. In part of the literature estimates of root production are based on statistically significant increments of root mass. Statistical significance, however, not only depends on the size of the difference but also on the sampling intensity. Although this method has been used in past decades, and the literature contains many estimates based on this method, a methodological study by Singh *et al.* (1984) showed that considerable <u>over</u> and <u>under</u> estimates can occur. For a full discussion of the biological and statistical aspects of the topic, including frequency of sampling in a range of systems, consult McClaugherty *et al.* (1982), Fairley and Alexander (1985), Goltz *et al.* (1984), Lauenroth *et al.* (1986), Hansson and Andren (1986), Singh *et al.* (1984) and Vogt *et al.* (1986). More reliable methods for estimating root turnover are based on separate study of root growth and root decay processes.

D.3.1 Root decay

Root decay (both on a root length and a root weight basis) can be studied in a modified litterbag method, incubating known amounts of excised roots (e.g. collected by sieving soil at harvest time) in a ceramic pot filled with sieved, root-free soil and placing the pots in the field (possibly at two depths). A screen cover on top of the pot may prevent soil fauna from removing root tissue. At regular intervals pots are retrieved and washed on a fine mesh sieve. Intact roots and root debris are collected separately.

D.3.2 In-growth cores

Seasonal patterns of root growth can be studied by the in-growth core method (Steen, 1984; Fabiao *et al.*, 1985). Soil from the depth at which the in-growth core will be placed is collected from the site, air-dried and roots removed by sieving. In-growth bags may be made from plastic sacking of a minimum of 4 mm mesh, or from more rigid polypropylene cylinders (4 to 10 cm diameter, sealed at the bottom and top with mesh). For each bag an amount of soil is weighed according to the required bulk density of the soil. Auger holes are prepared, and the sacks or cylinders are filled (*in situ*) with soil, compacting cm by cm, loosening the surface before new soil is added. The right procedure to obtain the required bulk density can be found by trial and error. At regular intervals (say 1 month) in-growth bags (minimum 3) are retrieved and washed over a fine mesh sieve. Data allow periods with rapid root growth to be identified.

Be ause of the altered physical, chemical and biological conditions in the sieved and repacked soil, growth of the roots into the bag may not accurately represent growth in the bulk soil, particularly on compacted or clay-rich soils. Periods of active root growth in different layers, however, should be accurately reflected.

The technique can also be used to record root response to localised fertiliser application or other heterogeneities in the soil (Cuevas and Medina, 1988; Hairiah *et al.*, 1991). Fertiliser application should be made by mixing the soil with a fertiliser solution prior to filling the bags.

D.3.3 Minirhizotrons

Simultaneous processes of root growth and root death or decay can only be quantified if the fate of individual roots can be followed. The simplest method uses a glass sheet against a soil profile wall, tracing roots on transparent polyethylene sheets. By using different colours of marking pens both new root growth and disappearance of roots can be quantified. An intermediate (but tedious) technique uses glass tubes inserted into the soil and regularly inspected (on grid lines in the glass) for roots using a mirror and a torch. More sophisticated versions use a fibreoptic system with a camera (Figure D.3) and flash light or a video system (Taylor, 1987).

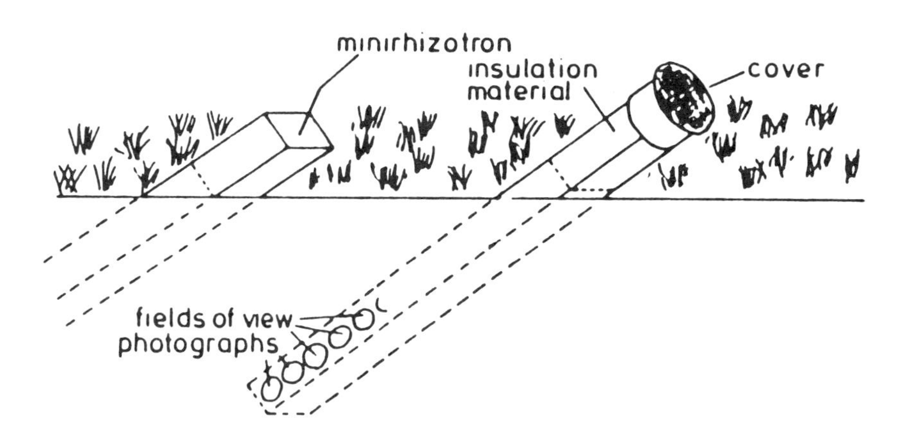

Figure D.3 Minirhizotron system to observe dynamics of root growth and decay under field conditions; at each time of observation a series of images with increasing depth is taken; these images can be analysed as shown in Figure D.4.

If a series of images has been obtained analysis can proceed as in Figure D.4: for each depth in the soil images are compared step by step (T1 with T2, T2 with T3 etc.) and the number of new root intersections with a grid is scored plus the number of intersections which has disappeared since the last observation. By adding up all new root intersections the actual root intensity on the observation plane can be expressed as a fraction of the total annual production. Curve fits (e.g. logistic) of root growth and root decay per layer can be obtained. If at one point in time, by destructive sampling, a reasonable estimate of standing root mass was obtained, these relative figures can be used to estimate annual root production. The assumption need not be made that root length on the observation surface has a known or constant relation to root length density in the soil. Calibration lines do in fact differ between soil horizons, crops, soil types etc. (Taylor, 1987). The main assumption needed for estimates of total root production are that the relative pattern of root growth and decay on the observation surface represents that in the soil. When glass or perspex (rigid) walls are used for the observation structure (mini-rhizotron), gaps between this surface and the soil may be unavoidable and roots grow and die in a "gap" environment. Gijsman *et al.*, (1991) described an inflatable minirhizotron system which reduces the problem of gap formation.

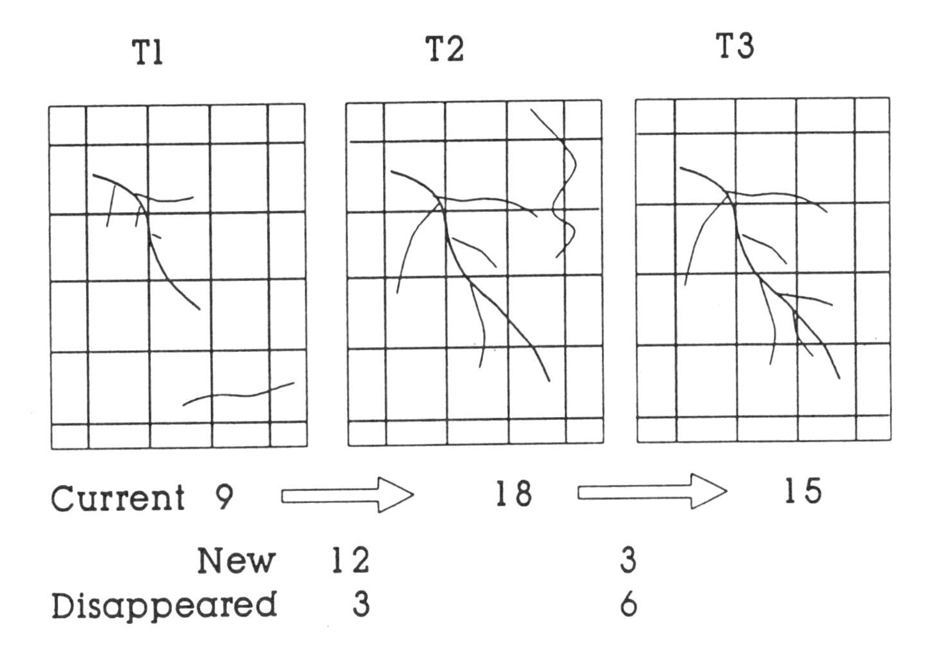

Figure D.4 Analysis of sequential images to derive root turnover.

D.4 ESTIMATE OF TOTAL CARBON INPUT TO THE SOIL

In addition to structural root tissue, carbon inputs to the soil include sloughed root cap cells, mucilage produced at the root tip, decayed root hairs and other cellular material, soluble carbohydrates, amino acids and other exudates, and CO_2 from root respiration. These root-rhizosphere transitions form a continuum, the study of which usually is based on ^{14}C labelling and sophisticated laboratory equipments.

References

Bohm, W. (1979) *Methods of Studying Root Systems.* Springer-Verlag, Berlin.

Bohm, W. and Kopke, U. (1977) Comparative root investigations with two profile wall methods. *Zeitschrift fur Acker-Pflanzenbau* **144**, 297-303.

Cahoon, G.A. and Morton, E.S. (1981) An apparatus for the quantitative separation of plant roots from soil. *Proceedings of the American Society for Horticultural Science* **78**, 593-596.

Caldwell, M.M., and Virginia, R.A. (1991) Root systems. In: Pearcy, R.W, Ehleringer, J., Mooney, H.A. and Rundel, R.W. (eds.), *Plant Physiological Ecology: Field Methods and Instrumentation.* Chapman and Hall, London, pp.367-398.

Cuevas, E. and Medina, E. (1988) Nutrient dynamics within Amazonian forests: Part II. Fine root growth, nutrient availability and litter decomposition. *Oecologia* **76**, 222-235.

Fabiao, A., Persson, H.A., and Steen, E. (1985) Growth dynamics of superficial roots in Portuguese plantations of *Eucalyptus globulus* Labill. studied with a mesh bag technique. *Plant and Soil* **83**, 233-242.

Fairley, R.I. and Alexander, I.J. (1985) Methods of calculating fine root production in forests. In: Fitter, A.H. (ed.), *Ecological Interactions in Soil.* Blackwell Scientific, Oxford, pp.37-42.

Floris, J. and Noordwijk, M. van (1984) Improved methods for the extraction of soil samples for root research. *Plant and Soil* **77**, 369-372.

Gholz, H.L., Hendry, L.C. and Cropper, W.P. (1986) Organic matter dynamics of fine roots in plantations of slash pine (*Pinus elliottii*) in north Florida. *Canadian Journal of Forestry Research* **16**, 529-538.

Gijsman, A.J., Floris, J. Noordwijk, M. van and Brouwer, G. (1991) An inflatable minirhizotron system for root observations with improved soil/tube contact. *Plant and Soil* **134**, 261-269.

Grzebisz W., Floris, J. and Noordwijk, M. van (1989) Loss of dry matter and cell contents from fibrous roots of sugar beet due to sampling, storage and washing. *Plant and Soil* **113**, 53-57

Hairiah K., Noordwijk, M. van and Setijono, S. (1991) Tolerance to acid soil conditions of the velvet beans Mucuna pruriens var. utilis and M. deeringiana. I. Root development. *Plant and Soil* **134**, 95-105.

Hansson, A.C. and Andren, O. (1986) Below-ground plant production in a perennial grass ley (*Festuca pratensis* Huds) assessed with different methods. *Journal of Applied Ecology* **23**, 657-666.

Hansson, A.C. and Steen, E. (1984) Methods of calculating root production and nitrogen uptake in an annual crop. *Swedish Journal of Agriculture Research* **14**, 191-200.

Kormanik, P.P. and McGraw, A.C. (1982) Quantification of vesicular-arbuscular mycorrhizae in plant roots. In: Schenck, N.C. (ed.), *Methods and Principles of Mycorrhizal Research.* American Phytopathological Society, St Paul, Minnesota, pp.37-45.

Lauenroth, W.K., Hunt, H.W., Swift, D.M. and Singh, J.S. (1986) Reply to Vogt *et al.* *Ecology* **67**, 580-582.

Mackie-Dawson, L.A. and Atkinson, D. (1991) Methodology for the study of roots in field experiments and the interpretation of results. In: Atkinson, D. (ed.), *Plant Root Growth, an Ecological Perspective.* Blackwell Scientific Publications, London, p.25-47.

McClaugherty, C.A., Aber, J.D. and Melillo, J.M. (1982) The role of fine roots in the organic matter and nitrogen budgets of two forested ecosystems. *Ecology* **63**, 1481-1491.

Newman, E.I. (1966) A method for estimating the total length of root in a sample. *Journal of Applied Ecology* **3**, 139-145.

Perrson, H. (1978) Root dynamics in a young Scots Pine stand in central Sweden. *Oikos* **30**, 508-519.

Phillips, J.M. and Hayman, D.S. (1970) Improved procedures for clearing and staining parasitic and vesicular arbuscular mycorrhizal fungi for rapid assessment of infection. *Transactions of the British Mycological Society* **55**, 158-161.

Richards, D., Goubran, F.H., Garwoli, W.N. and Daly, M.W. (1979) A machine for determining root length. *Plant and Soil* **32**, 69-76.

Rowse, H.R. and Phillips, D.A. (1974) An instrument for estimating the total length of root in a sample. *Journal of Applied Ecology* **11**, 309-314.

Schuurman, J.J. and Goedewaagen, M.A.J. (1971) *Methods for the Examination of Root Systems and Roots.* Second edition, Pudoc, Wageningen, the Netherlands, 86 pp.

Sieverding, E. (1991) *Vesicular-arbuscular mycorrhiza management.* GTZ publ. No. 224. Eschborn. ISBN 3-88085-462-9.

Singh, J.S., Lauenroth, W.K., Hunt, H.W. and Swift, D.M. (1984) Bias and random errors in estimators of net root production: a simulation approach. *Ecology* **65**, 1760-1764.

Smucker, A.J.M., McBurney, S.L. and Srivastava, A.K. (1982) Quantitative separation of roots from compacted soil profiles by the hydropneumatic elutriation system. *Agronomy Journal* **74**, 500-503.

Steen, E. (1984) Variation of root growth in a grass ley studied with a mesh bag technique. *Swedish Journal of Agricultural Research* **14**, 93-97.

Taylor, H.M. (ed.). (1987) *Minirhizotron Observation Tubes: Methods and Applications for Measuring Rhizosphere Dynamics.* American Society of Agronomy Special Publication 50, ASA, Madison, USA, 143 pp.

Taylor, H.M., Upchurch, D.R., Brown, J.M. and Rogers, H.H. (1991) Some methods of root investigations. In: McMichael, B.L. and Persson, H. (eds.), *Plant Roots and their Environment.* Elsevier Science Publishers, Amsterdam, p.553-564.

Tennant, D. (1975) A test of a modified line intersect method of estimating root length. *Journal of Ecology* **63**, 995-1001.

van Noordwijk, M. (1987) Methods for quantification of root distribution pattern and root dynamics in the field. *Proceedings of the 20th Colloquium of the International Potash Institute*, 243-262, Baden bie Wien.

van Noordwijk, M. and Brouwer, G. (1991) Review of quantitative root length data in agriculture. In: Persson, H. and McMichael, B.L. (eds.), *Plant Roots and their Environment.* Elsevier Science Publishers, Amsterdam, pp.515-525.

van Noordwijk, M. and Willigen, P de. (1991) Root functions in agricultural systems. In: Persson, H. and McMichael, B.L. (eds.), *Plant Roots and their Environment.* Elsevier Science Publishers, Amsterdam, pp.381-395.

van Noordwijk, M., Floris, J. and de Jager, A. (1985a) Sampling schemes for estimating root density in cropped fields. *Netherlands Journal of Agricultural Science* **33**, 241-262.

van Noordwijk, M., de Jager, A. and Floris, J. (1985b) A new dimension to observations in minirhizotrons: a stereoscopic view on root photographs. *Plant and Soil* **86**, 447-453.

van Noordwijk, M., Widianto, Heinen, M. and Hairiah, K. (1991a) Old tree root channels in acid soils in the humid tropics: important for crop root penetration, water infiltration and nitrogen management. *Plant and Soil* **134**, 37-44.

van Noordwijk, M., Kurniatun Hairiah, Syekhfani, M.S. and Flach, B. (1991) Peltophorum pterocarpa a tree with a root distribution suitable for alley cropping. In: Persson, H.

and McMichael, B.L. (eds.), *Plant Roots and their Environment.* Elsevier Science Publishers, Amsterdam, pp.526-532.

van Noordwijk, M., Brouwer, G. and Harmanny, K. (1992) Concepts and methods for studying interactions of roots and soil structure. *Geoderma* (in press).

Vogt, K.A., Grier, G.C., Gower, S.T., Sprugel, D.G. and Vogt, D.J. (1986) Overestimation of net root production: a real or imaginary problem? *Ecology* **67**, 577-579.

Ward, K.J., Klepper, B., Pickman, R.W. and Allmaras, R.R. (1978) Quantitative estimation of living wheat root lengths in soil cores. *Agronomy Journal* **70**, 675-677.

Welbank, P.J. and Williams, E.D. (1968) Root growth of a barley crop estimated by sampling with portable powered soil-coring equipment. *Journal of Applied Ecology* **5**, 477-481.

Wilhelm, W.W., Norman, J.M. and Newell, R.L. (1982) Semiautomated x-y-plotter-based method of measuring root length. *Agronomy Journal* **74**, 149-152.

Appendix E SOIL SOLUTION SAMPLING AND LYSIMETRY

Mike Hornung

Sampling systems are discussed under three main headings:

ISOLATED SOIL MASSES
TENSION/VACUUM/SUCTION SAMPLERS
TENSIONLESS COLLECTORS

Although the development of each method is outlined and the range of designs and construction materials described, specific recommendations are not made for TSBF projects since the choice of design often depends on site-specific constraints.

E.1 ISOLATED SOIL MASSES

Sampling systems of this type collect drainage water from a mass of soil enclosed in a container either in the laboratory or in the field. For field studies the drainage water is collected in suitable reservoirs situated in an adjacent trench. The systems can be divided into those in which the container is filled with disturbed, excavated soil and those which use undisturbed soil masses. The filled, or repacked, containers range from rectangular brick or concrete boxes, or tanks with surface areas of several squares metres down to repacked, circular cross-section metal or plastic tubes with diameters of a few centimetres. The containers should be constructed from inert materials to avoid solution-container interactions. The large *in situ* isolated block approach can only be used where there is a naturally impermeable subsoil.

Undisturbed masses range from a 10 m x 10 m *in situ* area isolated by vertical plastic sheets (Calder, 1976), to circular cross-section cores ca. 1 m in diameter in fibreglass (Belford, 1979) or metal (Roose and des Tureaux, 1970) containers, to cores in plastic pipes a few centimetres in diameter. The main constraint on the technique is the expense and/or technical difficulties of sampling undisturbed soil masses. The use of disturbed soil allows better replication of units than undisturbed soil but has the disadvantage that it is difficult to recreate the natural soil packing and structure. The larger types can accommodate a number of plants, which can be cropped, to provide a nutrient and water sink. Alternatively, Anderson *et al.* (1987) have used 0.5 m x 0.5 m x 0.3 m deep plastic trays filled with forest floor material with lengths of attached tree roots introduced to the lysimeters through ports in the sides of the trays.

The main advantages of using isolated soil masses are that a known volume and weight of soil is used, all drainage can be easily collected and, thus accurate element fluxes and budgets can be calculated. Disadvantages include the fact that only vertical drainage is possible (except with the large *in situ* masses), even when lateral drainage is an important factor on the undisturbed site and that roots from the soil are also excluded from the container; this is particularly important with the smaller columns.

E.2 TENSION/VACUUM/SUCTION SAMPLERS

E.2.1 Introduction

The extraction of soil solutions by applying a suction to a porous collector placed in the soil was first proposed by Briggs and McCall (1904). They used a Pasteur-Chamberlin filter tube connected to an evacuated two litre bottle. Since the exploratory study a wide variety of porous materials have been used in a number of designs (Table E.1) to produce a spectrum of sampling systems from the simple to the complex.

Table E.1 The main types of porous suction samplers

Ceramic cups (Wagner, 1962)
Ceramic tubes (Krone *et al.*, 1952; Czeratski, 1971)
Ceramic plates (Czeratski, 1959)
Alundum plates (Cole, 1958)
Acrylic plates (Bourgeois and Luvkulich, 1972)
Plastic cups (Quin and Forsythe, 1976)
Teflon rings (Morrison, 1982)
Teflon cups (Zimmerman *et al.*, 1978)
Sintered nickel cups (Hadrick *et al.*, 1977)
Cellulose-acetate hollow fibres (Jackson *et al.*, 1976)
Fritted glass tubes (Long, 1978)
Non-cellulosic hollow fibre tubing (Levin and Jackson, 1977)
Fritted glass plates (Chow, 1977)
Ceramic candles (Duke and Haise, 1973)

All the various systems, however, operate on the basic principles explored by Briggs and McCall (1904) - the porous sampler is installed in the ground, a vacuum is developed within the sampler and pore water is drawn into the sampler through the porous section when soil water suction is less than the applied vacuum.

E.2.2 Cup and ring-based sampling systems

The most widely used suction samplers have been ceramic cups. They are available in a variety of shapes and sizes and with a range of air-entry values. The most commonly used are ca. 6.0 cm in length, ca. 5.0 cm outside diameter and with an air-entry value of 1 bar. The precise composition of the ceramic material, and the pore sizes, vary from one manufacturer to another. The 1 bar air-entry value cups manufactured by Soilmoisture Equipment Corp. have approximately 2.9 μm pores. Small cups, or tubes, e.g. 6 mm x 65 mm, have been used to extract soil solutions in pot or glasshouse studies (e.g. Harris and Hansen, 1975).

Most ceramic cup sampler systems are based on Wagner's (1962) design. The cup is cemented to one end of a plastic tube, which usually acts as the sample reservoir. A bung seals the other end of the tube and a narrow bore tube through the bung serves to evacuate the sampler and to recover the water sample. A number of adaptions to this simple design have been reported. Two tubes are now usually incorporated through the bung; a short tube to evacuate the sampler and a longer one to recover the sample (e.g. Reeve and Doering, 1965). Other modifications are designed for sampling at considerable depth, (Parizek and Lane, 1970; Ward, 1973), to control sample volume (Chow, 1977), and to prevent flow of collected water during soil drying (Knighton and Stebflow, 1981). Other systems connect the sampler to additional samplers and/or vacuum reservoirs to increase the capacity and sampling period. A separate sample reservoir is particularly useful when the collector is positioned close to the soil surface and a long collector/sample tube assembly would be unstable.

Alundum, plastic, sintered nickel and teflon cup-based sampling systems are broadly similar in design to the above. Thus the cup is usually cemented or screwed to a tube which generally acts as the sample reservoir. More detailed accounts are contained in the references quoted in Table E.1. The teflon ring sampler introduced by Morrison (1982) provides an interesting variation on the existing sampler designs. A ring of porous teflon forms part of the sampler tube. The sample drains into the lower, sump-like section of the tube, thus preventing drainage back through the porous section when loss of vacuum or soil drying occurs.

Cup and ring-based samplers are usually installed vertically into a hole augered from the surface to the desired depth. It is recommended that the cup is pressed into puddled soil from the horizon being sampled, although it may be surrounded by silica sand. The cavity around and above the sampler is then backfilled, usually by reconstructing the profile. In order to prevent drainage outside the tube some workers have inserted a plug of clay around the tube but above the level of the cup or ring; this introduction of extraneous material may itself cause modifications to the soil water chemistry. Linden (1977) provides guidelines for both construction and installation of ceramic cup samplers. The installation procedures are applicable to other types of cup-based samplers.

Cup, and ring, type samplers can also be inserted horizontally into holes bored into the side wall of a pit. This will eliminate the problem of drainage down the outside of the sampler tube. Knighton and Steblow (1981) install their cup sampler into a cavity cut into the side wall of a pit. The cup is inserted into the upper face of the cavity which is then backfilled. As the authors suggest, this sampler design is useful near to the surface and when surface disturbance is likely.

E.2.3 Porous plate samplers

The first porous plate lysimeter reported was developed by Cole (1958) from 28 cm diameter alundum filter discs. In the original design, a rubber sheet was fastened to the edges of the disc; a connector valve was cemented into the rubber sheet and provided the outlet to the sample reservoir. A later design (Cole, 1968) had the alundum disc cemented to a plexiglass base, which incorporated an outlet pipe. Ceramic plate collectors are similar in design with the ceramic disc sealed to, or on, a base.

An interesting variation of the plate-based collector has been described by Bourgeois and Luvkulich (1972). The base of the collector is made from acrylic material, a rim around the upper surface of the base produces a shallow well which is infilled with a layer of silicon carbide powder. The authors claim that this material produces a better contact with the uneven soil face than the rigid alundum or ceramic plates.

Porous plate collectors are generally installed in small cavities cut into the side wall of a pit or trench. The plate is pressed against the upper face of the cavity which is then backfilled. The sample reservoir is separate from the actual collector and linked to it by a feeder tube.

The reservoir may be sited in the pit or trench excavated during installation, or on the ground surface. Cole (1968) describes a very sophisticated alundum plate-based system which links several plates to flow cells which monitor conductivity, acidity and rate of flow with output to a data logger.

E.2.4 Fritted glass tubes

Samplers of this type use fritted glass tubular filters. These are available in a number of sizes and with different porosities, e.g. Long (1978) used a 10 cm long filter with a 15.9 mm outside diameter. The filter is linked to a tube of some type which is then connected to a sample reservoir.

The samplers can be installed vertically or horizontally in a similar way to the cup and ring-based systems discussed earlier. Long (1978) suggests pressing the glass filter into a slurry made from the soil at the bottom of the hole, stressing the care needed because of the fragility of the fritted glass.

E.2.5 Hollow fibres

Silkworth and Grigal (1981) introduced their fibre bundle into the soil horizontally inside a polyvinylchloride (PVC) tube which had had "numerous" large diameter holes drilled into it. Levin and Jackson (1977) pushed a hollow steel tube horizontally into the soil; the bundle of fibres is then threaded through the tube and the tube then withdrawn to allow soil to collapse around the fibres.

E.2.6 Application of tension

As noted earlier, Briggs and McCall (1904) exerted a tension by evacuating a two litre bottle to which their sampler was connected. Evacuation of the sampling system is still the most commonly used method of applying the suction or tension. Hand or powered vacuum pumps may be used depending on the size and location of the system.

The volume of the sample and vacuum reservoir will determine the maximum sample volume which can be collected and the maximum duration of the vacuum. Additional "vacuum reservoirs" can be incorporated into the system to extend the sampling period (e.g. Cole, 1968). If power is available on site, a vacuum pump can be permanently connected to the system to provide a continuous suction.

A suction can also be applied by incorporating a "hanging column" or siphon arrangement into the system (Krone *et al.*, 1952; Cole, 1958). This will provide a continuous suction whenever the siphon column is intact.

E.2.7 Sample recovery

The solution sample may be recovered from the sample reservoir directly, by disconnecting the reservoir from the rest of the system, or by applying a suction or pressure to the sampler. Direct recovery can only be used when the sample reservoir is separate from the porous collector and accessible from the ground surface. When the sample reservoir is rigidly connected to the porous section and/or buried suction or pressure may be applied. Suction, or vacuum, recovery is used when the reservoir is relatively close to the surface whereas pressure is used for deeply buried collectors (Parizek and Lane, 1970; Morrison, 1982). Vacuum recovery is also frequently used to empty surface sited reservoirs so as to minimise disturbance to the system.

E.2.8 Solution-sampler interactions

The chemistry of the solution samples can be modified as a result of interactions with the sampler, mainly the porous material. A number of processes are probably involved, including leaching of elements from the porous material, sorption of solutes onto the sampler surface, screening of ions by the sampler and diffusion through the porous material (Hansen and Harris, 1975). Comparative studies involving two or more types of samplers (Zimmerman *et al.*, 1978; Silkworth and Grigal, 1981; Bottcher *et al.*, 1984) suggest that teflon samplers produce little modification of solution chemistry. Zimmerman *et al.* (1978) also examined teflon samplers and found no attenuation of a wide range of elements. Long (1978) also states that he found no contamination of samples collected with fritted glass tubes. Silkworth and Grigal, however report leaching of sodium from glass collectors.

Ceramic and alundum samplers, however, have been shown to release calcium, sodium, magnesium, silica and sulphur into solution while phosphate and nitrate concentrations are influenced by adsorption and ion screening (Hansen and Harris, 1975; Levin and Jackson, 1977). Trace metals such as Cu, Fe, Mn, Ni, Pb and Zn are also sorbed (Wolff, 1967; Stearns, 1980). The precise reactions will vary between cups made from different original clay mixes. Suitable preconditioning of ceramic and alundum samplers can eliminate or greatly reduce contamination by leaching impurities. Methods of pre-treatment are given by Grover and Lamborn (1970), Hetsch *et al.* (1979) and Neary and Tomassini (1985). These involve leaching with dilute acid, usually 0.1 M hydrochloric acid, prior to installation; Hetsch *et al.* (1979) follow this by leaching with large amounts of "equilibrium solution". After adequate preconditioning Hetsch *et al.* (1979) suggest that the ceramic cups they used could be used for studies of H, Na, K, NH_4, Ca, Mg, Mn, Al, S, Cl and NO_3. The importance, however of a prolonged equilibration period, after installation, must be stressed. Solution-sampler interactions continue to influence phosphate concentrations even after pre-treatment; indeed, Bottcher *et al.* (1984) found that acid leaching increased the amount of phosphate adsorption of alundum and ceramic materials. Nagpal (1982) has also shown an effect on potassium recovery from solutions and suggests sampling methods to minimise this. Sorption of trace metals also persists and may be enhanced. Cellulose-acetate hollow fibres also release a number of elements due to leaching (Silkworth and Grigal, 1981) e.g. Na. The adsorption characteristics of these materials do not appear to have been studied in detail.

E.2.9 Sampled soil volume and calculation of element fluxes

The volume of soil sampled by tension lysimeters cannot be defined precisely. It will vary with the design and type of the collector, the tension applied and the size and distribution of pores, channels and cavities in the soil. The very installation of the sampler results in soil disturbance which may well modify flow lines within the soil mass. Van der Ploeg and Beese (1977) modelled the flow of moisture towards suction units and concluded that even weak vacuums could produce large changes in seepage rate and that the radius of influence of a suction could be "several feet". Warrick and Amoozegar-Ford (1977) have produced expressions to enable one to assess the likely sampling volume and the region of influence. Further work of this type would seem to be needed, especially for different shapes of suction samplers. It should be noted, however, that Talsma *et al.* (1979) concluded that flow distribution around a porous cup was much closer to that of un-extracted soil than is predicted from soil water flow theory. They suggest that this is due to flow impedance near the cup wall which reduces cup uptake.

Much work with alundum and ceramic plate collectors has used relatively low tensions, ca. 0.1 bar; this could only "extract" water from the larger pores. In contrast, ceramic cups have frequently been used at tensions of 1 bar in a deliberate attempt to sample "capillary" water. Studies using plate-type samplers commonly assumed that the water was derived from a column of soil, stretching above the plate, with a diameter equal to that of the plate. Fluxes were calculated on the basis of this assumption; provided that a continuous suction was applied and sample collectors were adequate for the volumes involved. This approach seems questionable but may provide acceptable values with samplers sited near the surface. Cup, ring and tube shaped samplers should always be regarded as providing qualitative data. Separate data on soil-water contents and movement is required before element fluxes can be calculated. The suction samplers are, therefore, increasingly used in parallel with tensiometer or neutron probe systems. In sampling for flux calculations, it is preferable if a continuous suction is applied so that the chemistry is integrated through time.

E.2.10 Overview

The choice of tension/suction sampler system for use in any given study will depend on many factors including the aims of the study, soil characteristics, terrain, facilities available on site and the level of funding available. The following summarises some of the sampler characteristics which should be considered and gives relative costs of the samplers - whole system costs will depend on other considerations.

Ceramic cups: readily available, relatively cheap, fairly robust but brittle, vertical installation causes minimal disturbance, must be preconditioned to reduce contamination problems, after preconditioning suitable for major cations (although there are reservations about K), Cl, NO_3, SO_4 but not suitable for PO_4 or trace metals, qualitative only.

Alundum cups: similar to the above but not so readily available and hence more expensive.

Ceramic plates: more expensive than ceramic cups, thin plates rather fragile, installation more complex than with cup-based systems, similar contamination/adsorption problems to ceramic cups, some advantages in the plate design as it will only collect from a zone above the plate, semi-quantitative near surface.

Alundum plates: more expensive than ceramic cups but similar to ceramic discs, more disturbance of surrounding area of installation than with cup or ring-based systems, contamination/adsorption problems as above, plate design advantage as above, robust.

Acrylic plates: have to be constructed in house, costs mainly in labour - uncertain advantages of the plate design, improved contact with soil due to silicon carbide powder, solution interactions with the silicon carbide uncertain.

Plastic cups: relatively cheap, constructed in-house from available plastic sections, minimal installation disturbance of any systems, little contamination but there may be adsorption - no details available; only available with relatively large pores which may admit colloids, microbes etc. and lead to within sampler reactions.

Sintered nickel cups: moderately expensive, minimal disturbance of all cup systems, unstable in acid soils, interactions with solutions limits use to a few elements.

Teflon cups: relatively expensive, have to be made in-house, little disturbance at installation, no known contamination problems, robust.

Teflon rings: commercially available, relatively expensive, little disturbance at installation, may be problems with blocking of pores, no known contamination problems, robust.

E.3 TENSIONLESS COLLECTORS

E.3.1 Introduction

Tensionless collectors are designed to intercept and collect water draining freely, vertically or obliquely, through soil. Probably the earliest use of this type of device was by Ebermeyer (1876) in his classic study of nutrient transfers in German forests; he used a trough-like collector to sample water draining from the forest floor. Since this pioneering work a wide variety of shapes and sizes of collector have been constructed from a range of materials. Thus copper, stainless steel, galvanised steel, perspex, polyvinylchloride and polypropylene have all been used. The collectors are generally made in-house and may be constructed from sheets of the selected material or may adapt and incorporate preformed units, e.g. boxes, bowls, trays, funnels or rainwater guttering. The devices are usually installed into the sidewall of a pit or trench. The collector may be connected directly to the sample reservoir or more commonly, linked to it by a feeder tube.

E.3.2 Trough, box and funnel-based collectors

Such collectors are generally installed in a cavity excavated in the sidewall of a pit or trench. The device may be filled with soil excavated from the cavity before it is inserted and brought into contact with the upper face of the cavity. Alternatively the collector is pushed upwards, into the upper face of the cavity. The approach chosen will depend on factors such as soil texture and stoniness, and density and size of roots. As in all tensionless collectors, a filter system should be incorporated to prevent clogging of tubes leading to the sample collector. In filled bowls, troughs etc., a layer of acid wash sand or plastic beads can be placed in the base before filling. After insertion of the collector, the cavity should be backfilled to keep the device in place and in contact with the overlying soil.

Jordan (1968) has described a trough-shaped device which has proved useful in a wide range of site types. The trough is constructed from stainless steel and rests upon adjustable wooden supports. Once inserted into the sidewall cavity the supports can be adjusted in length in order to press the collector against the upper face of the cavity. A fibreglass screen lined with glass wool is suspended inside the trough. Steel rods run along the underside of the fibreglass screen to the outlet tube. The glass wool, fibreglass and steel rods together help to overcome surface tension within the soil and ensure movement of water from the soil into the collector. As Jordan notes, without some such device water tends to hang on the soil face rather than drip into the collector.

Before the collector is installed into the pre-dug cavity soil is tamped on top of the glass wool-fibreglass screen until it is level with the top of the trough. The design can be copied using preformed plastic guttering, nylon bolting cloth for the screen and plastic rods below the screen.

Boerner (1982) has described a simple trough shaped tension-free lysimeter for use in sandy soils. The collector is a V-shaped, made from 6 mm thick plexiglass, with sides 50 cm x 7.5 cm. It was installed into tunnels, cut into the end walls of soil pits, at an angle of ca. 20° to the horizontal. After inserting the trough the tunnel was repacked with the excavated soil. A "sump plate" inserted vertically into the exposed end of the trough prevents soil slumping into the exposed end of the trough. A "dust cover" prevents extraneous material entering the exposed section. The sample bottle is housed on the floor of the soil pit and linked to the lysimeter with flexible plastic tube.

E.3.3 Sheet or tray-based collectors

A frequently used type of collector comprises a sheet or tray-like device inserted horizontally, or at a slight angle to ensure drainage, into the side wall of a pit. A popular design has a flat base with a small front edge, side walls which reduce in height from front to back but no vertical rear edge.

Parizek and Lane (1970) refer to this type of collector as a pan lysimeter. They constructed 30 cm x 42 cm pan lysimeters from 16 gauge galvanised metal. A copper tube was soldered into the front, raised edge of the pan to allow the water to drain away to a sample reservoir. The pans were installed into the side wall of a trench. A sheet metal blade was first driven into the trench face to provide an access slit into which the pans were pushed. Wooden sides were put into the trench with a small gap between them and the pit face, and then this gap was backfilled with soil. The sample bottles, linked to the pan lysimeters, were placed on the trench floor.

Nys (personal communication), has constructed a sampler of this type from 1 cm polypropylene sheet; the sections are joined by heat welding. The resulting collector is very robust and, with the leading edge chamferred, will cut through roots up to ca. 1.0 cm diameter during installation. The collectors were forced into the pit face using hydraulic jacks. A small quantity of acid washed sand was placed above the outlet pipe to prevent clogging. After forcing the plate into the trench face a narrow sheet of plastic is inserted above the exposed face of the plate to prevent drainage down the soil face into the collector. To maintain soil temperature regimes etc., Nys places a wooded wall a short distance out from the trench face and backfills the gap by reconstructing the profile.

Reynolds (personal communication) has constructed a similar, but less robust device from 3 mm PVC sheet. The cut sections are cemented together using PVC cement. This collector is not strong enough to be forced into the pit face; a cut is therefore made with a knife or trowel before insertion.

Large plastic sheets can be inserted into a pit face, at a slight angle to ensure drainage, with a small edge left protruding from the face. The edge overhangs a trough or gutter which runs along the face and is connected to a sample bottle. This type of collector is referred to again below. Trough-like containers can also be inserted into a pit face rather than into a cavity cut into that face. The trough should have no vertical rear end piece. This type of collector can be constructed from plastic gutters. As the plastic material is not very robust, a cut may have to be made into the soil face before insertion. Alternatively, a metal former can be driven in first to ease insertion of the plastic trough.

The direct insertion of these pan and trough types of collector into a pit face can result in a smearing of the soil face in contact with the plate or trough. This may effectively seal the soil, greatly reducing the yield of water and undermine any calculation of fluxes. The risk of smearing is reduced if the collector is inserted during dry weather but the texture of the soil is the main determining factor. Direct insertion is also difficult in stony soils or where there are abundant large roots.

E.3.4 Tensionless collectors on sloping sites

Most tensionless lysimeters have been used on level or gently sloping sites and have been assumed to collect water draining vertically through the soil mass. On sloping sites, however, a large part of the water movement is lateral or oblique. Parisek and Lane (1970) used pan lysimeters to intercept this laterally moving water. An alternative approach uses a collection system first used in hydrological studies on slopes (Whipkey, 1965) and probably first used in solute flux studies by Roose (1968). The approach uses a trench dug at right angles to the slope. Sheets of plastic or metal are pushed into the upslope face of the pit, commonly at the base of each main soil horizon, so that little of the sheet still protrudes from the trench face. Gutters are installed along the face of the pit under each sheet overhang. The sheets intercept the laterally moving water which moves across their upper surface and drips down into the gutter.

At one end of each gutter is a connector and a feeder pipe which links it to a sample bottle on the floor of the trench. The gutters are usually mounted onto a wooden frame and wooden walls may be constructed to prevent collapse of the pit faces. Detailed descriptions, and/or diagrams, of this type of installation are provided by Roose (1968), Knapp (1973) and Williams *et al.* (1984).

E.3.5 Sample contamination

Sample contamination due to solution-sampler interactions can be virtually eliminated by careful selection of construction materials. Metals, other than stainless steel, should probably be avoided; they are gradually attacked, especially in acid soils. Plastic materials are particularly suitable but tests should be carried out on any new plastic to assess the possible adsorption of phosphate or trace metals and the release of fillers. Some plastics also "age" relatively rapidly and their adsorption characteristics may then change. In general, "high density" plastics have much lower adsorption than low density ones. The use of fibreglass filters, or screens, and of silica sand to avoid clogging of drainage pipes may lead to silica enhancements in solutions. The fibreglass may also have undesirable adsorption characteristics.

E.3.6 Calculation of fluxes

When tensionless lysimeters are used in solute flux studies on level sites, it is assumed that they collect water draining from a column of soil, with a cross-sectional area equal to that of the lysimeter, stretching vertically from the lysimeter to the ground surface. This may be a valid assumption on some sites but on others routing of water down macro-pores may concentrate, or disperse, flow. As a result, closely adjacent collectors may consistently collect very different volumes (e.g. Russel and Ewel, 1985). Flux studies should, therefore, use as large a number of lysimeters as possible.

On sloping sites, it is even more difficult to define the source area of the water collected. One approach is to isolate an area stretching upslope from the trench housing the collectors. This can be done by driving sheets of material vertically into the soil or excavating a boundary trench, inserting a vertical, waterproof barrier and backfilling the trench on either side of the barrier.

In any study of fluxes, it is essential that the sample reservoirs are large enough to accommodate all the water draining from the lysimeter during the sampling interval, or that a sample splitter is incorporated. The excavation of the pit or trench prior to installation of the tensionless lysimeter will cause alterations to drainage lines within the soil. If a pit, or trench, is left open it will have a sump effect with drainage lines converging on the pit, especially on sloping ground. Backfilling the pit will reduce this effect but not eliminate it. The magnitude of the effect will vary with soil texture and other site and soil characteristics.

References

Anderson, J.M., Leonard, M.A. and Ineson, P. (1987) Lysimeters with and without tree roots for investigating the role of macrofauna in forest soils. In: Harrison, A.F., Ineson, P., Heal, O.W. (eds.), *Field Methods in Terrestrial Ecosystem Nutrient Cycling.* Elsevier, Amsterdam.

Belford, R.K. (1979) Collection and evaluation of large soil monoliths for soil and crop studies. *Journal of Soil Science* **30**, 363-373.

Boerner, R.E.J. (1982) An inexpensive, tension-free lysimeter for use in sandy soils. *Bulletin of the Torrey Botanical Club* **109**, 80-83.

Bottcher, A,B., Miller, L.W. and Cambell, K.L. (1984) Phosphorus adsorption in various soil-water extraction cup materials: Effect of acid wash. *Soil Science* **137**, 239-245.

Bourgeois, W.W. and Lavkulich, L.M. (1972) A study of forest soils and leachates on sloping topography using a tension lysimeter. *Canadian Journal of Soil Science* **52**, 375-391.

Briggs, L.J. and McCall, A.F. (1904) An artificial root for inducing capillary movement of soil moisture. *Science* **20**, 566-569.

Calder, I.R. (1976) The measurement of water losses from forested area using a "natural" lysimeter. *Journal of Hydrology* **31**, 311-325.

Chow, T.L. (1977) A porous cup soil-water sampler with volume control. *Soil Science* **124**, 173-176.

Cole, D.W. (1958) Alundum tension lysimeter. *Soil Science* **85**, 293-296.

Cole, D.W. (1968) A system for measuring conductivity, acidity and rate of water flow in a forest soil. *Water Resources Board* **4**, 1127-1136.

Czeratzki, W. (1959) Untersuchung der Wasserbewegung im Boden mit Hilfe von Unterdrucklysimeter. *Z. Pflanzenern. Düngung. Bodenkunde* **87**, 223-229.

Czeratzki, W. (1971) Saugvorrichtung für kapillar gebundene Bodenwasser. Landbauforschg. *Völkenrode* **21**, 13-14.

Duke, H.R. and Haise, H.R. (1973) Vacuum extractors to assess deep percolation losses and chemical constituents of soil water. *Soil Science Society of America Proceedings* **37**, 963-964.

Ebermayer, E. (1878) *Die gesammte Lehre der Waldstreu mit Rücksicht auf die chemische Statik des Waldbaues.* Springer, Berlin.

Grover, B.L. and Lamborn, R.E. (1970) Preparation of porous ceramic cups to be used for extraction of soil water having low solute concentrations. *Soil Science Society of America Proceedings* **34**, 706-708.

Hansen, E.A. and Harris, A.R. (1975) Validity of soil-water samples collected with porous ceramic cups. *Soil Science Society of America Proceedings* **39**, 528-536.

Harris, A.R. and Hansen, E.A. (1975) A new ceramic cup soil-water sampler. *Soil Science Society of America Proceedings* 39, 157-158.

Hetsch, W., Beese, F. and Ulrich, B. (1979) Die Beeinflussung der Bodenlösung durch Saugkerzen aus Ni-Sintermetall und Keramick. Z. Planzenernaehr. *Bodenkunde* **142**, 29-38.

Jackson, D.R., Brinkley, R.S. and Bondietti, E.A. (1976) Extraction of soil water using cellulose-acetate hollow fibres. *Soil Science Society of America Proceedings* **40**, 327-329.

Jordan, C.F. (1968) A simple, tension-free lysimeter. *Soil Science* **105**, 81-86.

Knapp, B.J. (1973) *A system for the field measurement of soil water movement.* British Geomorphological Research Group Technical Bulletin Number 9.

Knighton, M.D. and Streblow, D.W. (1981) A more versatile soil water sampler. *Soil Science Society of America Journal* **45**, 158-159.

Krone, R.B., Ludwig, H.F. and Thomas, J.F. (1952) Porous tube device for sampling soil solutions during water-spreading operations. *Soil Science* **73**, 211-219.

Levin, M.J. and Jackson, D.R. (1977) A comparison of *in situ* extractors for sampling soil water. *Soil Science Society of America Journal* **41**, 535-536.

Linden, D.R. (1977) *Design, installation and use of porous ceramic samplers for monitoring soil-water quality.* USDA Technical Bulletin Number 1562, 11 pp.

Long F.J. (1978) A glass filter soil solution sampler. *Soil Science Society of America Journal* **42**, 834-835.

Morrison, R.D.C. (1982) A modified vacuum-pressure lysimeter for soil water sampling. *Soil Science* **134**, 206-210.

Nagpal, N.K. (1982) Comparison among and evaluation of ceramic porous cup soil water samples for nutrient transport studies. *Canadian Journal of Soil Science* **62**, 685-694.

Neary, A.J. and Tomassini, F. (1985) Preparation of alundum/ceramic plate tension lysimeters for soil water collection. *Journal of Soil Science* **65**, 169-179.

Parizek, R.R. and Lane, B.E. (1970) Soil-water sampling using pan and deep pressure-vacuum lysimeter. *Journal of Hydrology* **11**, 1-21.

Quin, B.F. and Forsythe, L.J. (1976) All-plastic suction lysimeters for the rapid sampling of percolating soil water. *New Zealand Journal of Science* **19**, 145-148.

Reeve, R.C. and Doering, E.J. (1965) Sampling the soil solution for salinity appraisal. *Soil Science* **99**, 339-344.

Roose, E.J. (1968) Un dispositif de mesure du lessivage oblique dans les sols en place. Cah. ORSTOM, ser. *Pedologie* **VI**, 235-249.

Roose, E.J. and Des Tureaux, M. (1970) Deux methodes de nature du drainage vertical dans les sols en place. *Agronomique Tropicale* **25**, 1079-1087.

Russel, A.E. and Ewel, J.J. (1985) Leaching from a tropical andept during big storms, a comparison of three methods. *Soil Science* **139**, 181-189.

Silkworth, D.R. and Grigal, D.F. (1981) Field comparison of soil solution samplers. *Soil Science Society of America Journal* **45**, 440-442.

Talsma, T., Hallam, P.M. and Mansell, R.S. (1979) Evaluation of porous cup soil-water extractors, physical factors. *Australian Journal of Soil Resources* **17**, 417-422.

van der Ploeg, R.R. and Beese, F. (1977) Model calculations for the extraction of soil water by ceramic cups and plates. *Soil Science Society of America Journal* **41**, 466-470.

Wagner, G.H. (1962) Use of porous ceramic cups to sample soil water within the profile. *Soil Science* **94**, 379-386.

Warrick, A.W. and Amoozegar-Fard, A. (1977) Soil water regimes near porous cup water samplers. *Water Resources Research* **13**, 203-207.

Whipkey, R.Z. (1965) *Measuring subsurface storm flow from simulated rainstorms - a plot technique.* U.S. Forest Service Central States Forest Experimental Station Research Note CS-29.

Williams, A.F., Ternan, J.L. and Kent, M. (1984) Hydrochemical characteristics of a Dartmoor hillslope. In: Burt, T.P. and Walling, D.E. (eds.), *Catchment Experiments in Fluvial Geomorphology*. Geobooks, Norwich, U.K.

Wolff, R. (1967) Weathering of woodstock granite near Baltimore, Maryland. *American Journal of Science* **265**, 106-117.

Zimmermann, C.F., Price, M.T. and Montgomery, J.R. (1978) A comparison of ceramic and teflon *in situ* samplers for nutrient pore water determinations. *Estuarine Coastal Marine Science* **6**, 1-5.

Appendix F NITROGEN AVAILABILITY

Cheryl Palm, Phil Robertson and Peter Vitousek

F.1 INTRODUCTION

Nitrogen is undoubtedly the most intensively studied plant nutrient in both natural and managed ecosystems. Despite its importance to plant growth and ecosystem productivity, there is still no reliable soil test for nitrogen availability (Stanford, 1982). The reasons for this are numerous (see Stanford, 1982 and Stevenson, 1986 for reviews).

Soil nitrogen for the most part is in the organic form (Stevenson, 1986) and unavailable to plants. Organic nitrogen is mineralised to ammonium, a plant-available form, by microbial processes. The rate of mineralisation depends on many factors including temperature and rainfall, the quality of the soil organic nitrogen, and the quality of organic inputs to the system. Once in the mineral form, nitrogen, like other plant nutrients, can be taken up by plants or microbes (immobilisation) or lost from the system by leaching. In addition, nitrogen can undergo various transformations, many of them biologically mediated. Many of these transformations such as nitrification, volatilisation, and denitrification can ultimately lead to losses from the system.

In natural ecosystems nitrogen losses are generally assumed to be small because of efficient nutrient cycling, mineralisation occurring in synchrony with plant uptake. In managed systems however, this synchrony may be disrupted leading to a build up of mineral nitrogen in the soil.

This nitrogen is susceptible to losses via the various pathways discussed. Leaching losses can be large, especially in the humid tropics where rainfall is high and a majority of the soils have little retention capacity for cations or anions. Losses of nitrogen through denitrification may be as large, or larger, than through leaching in disturbed or managed systems. Little is known, however, about denitrification in tropical ecosystems, natural or managed. The potential for denitrification should be high in tropical systems where nitrification rates are high and high rainfall can lead to anaerobic conditions in the soil. There are indications that denitrification is higher in undisturbed tropical forest ecosystems than those in temperate regions (Keller *et al.*, 1986; Mooney *et al.*, 1987) and that these rates increase with disturbance of the system (Keller *et al.*, 1986; Matson *et al.*, 1987). In arid tropical ecosystems, most nitrogen loss occurs as a result of fires.

One of the major objectives of the TSBF Programme is to find ways of minimising losses of nitrogen (as well as other nutrients) in managed systems by synchronising mineralisation with plant uptake. This will require an understanding of the controls of the various transformations and fluxes of nitrogen and how these are affected by management. Part of the problem in understanding nitrogen cycling in ecosystems has been the lack of methodologies for measuring the different processes. In this section methodologies for estimating net mineralisation and denitrification and the problems associated with them are presented. The collection of leachate is discussed in Appendix E, "Soil solution sampling and lysimetry".

F.2 MINERALISATION

The concentration of mineral nitrogen in soil reflects a balance between inputs through mineralisation of organic N and outputs through microbial immobilisation, plant uptake, leaching, and denitrification. Most methods for estimating nitrogen availability assume it to be equivalent to net nitrogen mineralisation (gross mineralisation minus immobilisation); they are designed to prevent (or quantify) root uptake, leaching, and denitrification by incubating soil in the absence of active roots. Any method however, which involves either severing roots or removing them from soil alters the soil environment to a greater or lesser extent. Hence, any such effort to measure soil nitrogen availability interferes with the process, and a broad-scale measurement programme must seek a comparable and realistic index of nitrogen availability rather than a direct measurement of the process.

No single method for estimating nitrogen availability has gained universal acceptance, and indeed it is unlikely that any single method will prove applicable to all sites and conditions (Keeney, 1980). Moreover, relatively few of the extant methods have been tested in tropical soils. The TSBF Programme initially adopted a field incubation method (below) as the standard, but workers in high-rainfall areas have reported problems with flooding of soil cores. Other studies have reported substantial losses of field- incubated cores to soil fauna (Vitousek and Denslow, 1986). It was therefore decided to use aerobic incubations in the laboratory as the simplest standard method, and to suggest optional use of more realistic field incubations where practical (see (b) below).

Laboratory methods for estimating N-mineralisation were reviewed by Brown (1982), who showed that a wide range of incubation times, temperatures and conditions, and extraction procedures have been employed with little attempt at standardisation. Many authors have incubated soil samples over long time periods assuming that mineralisation is linear over time. Myers (1975) however, reported for tropical soils that ammonification rates in the first seven days of incubation were about six times higher than between 14 and 28 days.

It is also difficult to control soil moisture while maintaining aerobic conditions over long periods of incubation. Quartz sand may be used to disperse soil, but ambient field moisture conditions cannot be maintained using this method. As a compromise on the various limitations of laboratory incubations, the simplest aerobic incubation procedure will be used as a baseline for TSBF studies (see Section 6.7.3 for details). Details are also given (Brown, 1982) for an anaerobic (waterlogged) laboratory incubation method of Waring and Bremner (1964). This method is rapid, and it may have potential as an index of the labile pool of plant-available N; it does not approximate field mineralisation processes in any way.

Methods used to estimate patterns of N mineralisation under field conditions include:

a) Incubation of disturbed (sieved or mixed) soil in plastic bags buried in the field (e.g. Westermann and Crowthers, 1980).

b) Incubation of relatively undisturbed soil columns (cores) enclosed in plastic bags or columns under field conditions (Nadelhoffer *et al.*, 1985; Matson *et al.*, 1986; Raison *et al.*, 1987). Rapp *et al.* (1979) performed *in situ* incubations after isolating the soil within thin metal cans pushed into the soil.

c) Measurement of mineral-N collected by ion exchange resins placed in the field for extended periods (e.g. Hart and Binkley, 1985).

d) Determining the effects of varying temperature and moisture on N mineralisation in disturbed soils in the laboratory, and then assuming that these relationships can be linked to variations in these parameters in the field - i.e. a modelling approach (e.g. Cameron and Kowalenko, 1976; Marion *et al.*, 1981; Macduff and White, 1985).

In addition to changes induced by severing or removing roots (Hendrickson and Robinson, 1984) the following potential difficulties may apply:

i) soil disturbance markedly affects mineralisation (e.g. Runge, 1974; Nordmeyer and Richter, 1985). This effect will be significant for methods (a) and (d); it may be quite small for method (b);

ii) rapid fluctuations in soil water content markedly affect mineralisation in many environments. With methods (a) and (b) the soil moisture content at the time of sampling is maintained throughout the incubation period;

iii) method (c) estimates the amount of nitrogen in percolating water (Binkley, 1984), not net nitrogen mineralisation. It is insensitive to any N either absorbed by roots or exchanged on soil surfaces; and

iv) for modelling approaches to be useful, temperature and moisture response surfaces must be determined for undisturbed soils, and account needs to be taken of seasonal variations in the pools of mineralisable organic N in soils (Popovic, 1971; Ellis, 1974; Theodorou and Bowen, 1983; Nordmeyer and Richter, 1985).

Of the above methods, (b) appears most suitable for a wide range of sites.

F.3 DENITRIFICATION

Denitrification is the biological reduction of soil nitrate to nitrous oxide (N_2O) and dinitrogen (N_2) gases. The process is carried out by a diverse group of bacteria using nitrogen oxides as terminal electron acceptors in lieu of oxygen under anaerobic conditions. In well-aerated soils these conditions most often exist within soil aggregates or where high concentrations of organic matter create strong, localised O_2 sinks.

The reductive pathway for denitrification is generally accepted to be:

NO_3^- --------> NO_2^- --------> NO --------> N_2O --------> N_2

with the ratio of $N_2O:N_2$ liberated by the denitrifying community highly variable, under the control of a wide range of environmental factors (Firestone *et al.*, 1980). Though denitrification is in most soils the largest single source of nitrogen gas, N_2O may additionally arise by a variety of other pathways. These include the chemical decomposition of hydroxylamine and nitrite reduction during nitrification (Blackmer *et al.*, 1980; cf. Poth and Focht, 1985), dissimilatory nitrite and nitrate reduction to ammonium (Bleakley and Tiedje, 1982), microbial nitrate assimilation (Satoh *et al.*, 1981), and other pathways as yet poorly described (Robertson and Tiedje, 1987).

Experimental methods for quantifying denitrification are reviewed by Focht (1982), Tiedje (1982), and Tiedje *et al.* (1989), including ^{15}N tracer techniques. The simplest way to measure denitrification in field studies is the acetylene inhibition technique (Duxbury, 1986; Tiedje *et al.*, 1989). Acetylene at relatively low concentrations (<10 kPa or <10% v/v at 1 atmosphere) blocks the expression of N_2O reductase in most soils, so that the N_2O which might otherwise have been reduced to N_2 remains as N_2O, easily measured against background atmospheric concentrations.

Ryden and Dawson (1982) report a field method where a constant stream of acetylene gas was injected 1 m below a pair of boxes covering a pasture sward while a second pair of boxes was untreated. Gas products from the two series of chambers were collected over 3-4 hr and scrubbed for N_2O using a molecular sieve; N_2O was then recovered and determined by gas-liquid chromatography. The estimates of N losses by denitrification using this method at intervals over the growing season were in close agreement with estimates obtained from N budgets for the pasture system.

The main limitations on the acetylene inhibition technique are ensuring (i) that acetylene diffuses to the soil microsites where denitrification is taking place, and (ii) since acetylene also inhibits nitrification, that denitrification rates are quantified before nitrate becomes limiting. Both chamber and soil core techniques have been used to quantify denitrification using the acetylene inhibition method; both techniques yield similar results (Ryden and Skinner, 1987; Tiedje *et al.*, 1989) but the static core technique has the advantages of being relatively inexpensive and amenable to large sample sizes, important considerations given the high degree of spatial heterogeneity typical for this process in a variety of environments. The static core's principal disadvantage is that great care must be taken during sampling and subsequent analysis to avoid disrupting anaerobic microsites in soil by disrupting soil structure. The recommended TSBF procedure is given in Section 6.8.2.

References

Binkley, D. (1984) Ion-exchange resin bags for assessing soil N availability: the importance of ion concentration, water regime, and microbial competition. *Soil Science Society of America Journal* **48**, 1181-1184.

Blackmer, A.M., Bremner, J.M. and Schmidt E.L. (1980) Production of N_2O by ammonia-oxidising chemoautotrophic microorganisms in soil. *Applied and Environmental Microbiology* **40**, 1060-1066.

Bleakley, B.H. and Tiedje, J.M. (1982) Nitrous oxide production by organisms other than nitrifiers or denitrifiers. *Applied and Environmental Microbiology* **44**, 1342-1348.

Brown, C.M. (1982) Nitrogen mineralisation in soils and sediments. In: Burns, R.G. and Slayter, J.H. (eds.), *Experimental Microbial Ecology.* Blackwell Scientific Publications, Oxford, pp.154-163.

Cameron, D.R. and Kowalenko, C.G. (1976) Modelling nitrogen processes in soil: mathematical development and relationships. *Canadian Journal of Soil Science* **56**, 71-78.

Duxbury, J.M. (1986) Advantages of the acetylene method for measuring denitrification. In: Hauck, R.D. and Weaver, R.W. (eds.), *Field Measurement of Dinitrogen Fixation and Denitrification.* Soil Science Society of America, Madison, Wisconsin, pp.73-91.

Ellis, R.C. (1974) The seasonal pattern of nitrogen and carbon mineralisation in forest and pasture soils in southern Ontario. *Canadian Journal of Soil Science* **54**, 15-28.

Firestone, M.K., Firestone, R.B. and Tiedje, J.M. (1980) Nitrous oxide from soil denitrification: factors controlling its biological reduction. *Science* **208**, 749-751.

Focht, D.D. (1982) Denitrification. In: Burns, R. and Slayter, J.H. (eds.), *Experimental Microbial Ecology.* Blackwell Scientific, Oxford, pp.194-211.

Hart, S.C. and Binkley, D. (1985) Correlations among indices of forest soil nutrient availability in fertilised and unfertilised loblolly pine plantations. *Plant and Soil* **85**, 11-21.

Hendrickson, O.Q. and Robinson, J.B. (1984) Effects of roots and litter on mineralisation processes in forest soil. *Plant and Soil* **80**, 119-140.

Keeney, D.R. (1980) Prediction of soil nitrogen availability in forest ecosystems: a literature review. *Forest Science* **26**, 159-171.

Keller, M., Kaplan, W.A. and Wofsy, S.C. (1986) Emission of N_2O, CH_4, and CO_2 from tropical forest soils. *Journal of Geophysical Research* **91**, 11791-11802.

Macduff, J.H. and White, R.E. (1985) Net mineralisation and nitrification rates in a clay soil measured and predicted in permanent grassland from soil temperature and moisture content. *Plant and Soil* **86**, 151-172.

Marion, G.M., Kummerow, H. and Miller, P.C. (1981) Predicting nitrogen mineralisation in chaparral soils. *Soil Science Society of America Journal* **45**, 956-61.

Matson, P.A., Vitousek, P.M., Ewel, J.J., Mazzarino, M.H. and Robertson, G.P. (1986) Nitrogen transformations following tropical forest falling and burning on a volcanic soil. *Ecology* **68**, 491-502.

Mooney, H.A., Vitousek, P.M. and Matson, P.A. (1987) Exchange of materials between terrestrial ecosystems and the atmosphere. *Science* **238**, 926-932.

Myers, R.J.K. (1975) Temperature effects on ammonification and nitrification in a tropical soil. *Soil Biology and Biochemistry* **7**, 83-86.

Nadelhoffer, K.J., Aber, J.D. and Melillo, J.M. (1985) Fine roots, net primary production, and soil nitrogen availability: a new hypothesis. *Ecology* **66**, 1377-90.

Nordmeyer, H. and Richter, J. (1985) Incubation experiments on nitrogen mineralisation in loess and sandy soils. *Plant and Soil* **83**, 433-45.

Popovic, B. (1971) Effect of sampling data on nitrogen mobilisation during incubation experiments. *Plant and Soil* **34**, 381-92.

Poth, M. and Focht, D.D. (1985) ^{15}N kinetic analysis of N_2O production by Nitrosomonas europaea: an examination of nitrifier denitrification. *Applied and Environmental Microbiology* **49**, 1134-1141.

Raison, R.J., Connell, M.H. and Khanna, P.K. (1987) Methodology for studying fluxes of soil mineral-N *in situ. Soil Biology and Biochemistry* **19**, 521-530.

Rapp, M., Leclerc, M. Cl. and Lossaint, P. (1979) The nitrogen economy of a Pinus pinea L. stand. *Forest Ecology and Management* **2**, 221-231.

Robertson, G.P. and Tiedje, J.M. (1987) Nitrous oxide sources inaerobic soils: nitrification, denitrification and other biological processes. *Soil Biology and Biochemistry* **19**, 187-193.

Ryden, J.C. and Dawson, K.P. (1982) Evaluation of the acetylene-inhibition technique for the measurement of denitrification in grassland soils. *Journal of the Science of Food and Agriculture* **33**, 1197-1206.

Ryden, J.C. and Skinner, J.H. (1987) Soil core incubation system for the field measurement of denitrification using acetylene-inhibition. *Soil Biology and Biochemistry* **19**, 753-757.

Satoh, T.H., Hom, S.S.M. and Shanmugam, K.T. (1981) Production of nitrous oxide as a product of nitrite metabolism by enteric bacteria. In: Lyons, J.M., Valentine, R.C., Phillips, D.A., Rains, D.W. and Huffaker, R.C. (eds.), *Genetic Engineering of Symbiotic Nitrogen Fixation and Conservation of Fixed Nitrogen.* Plenum Press, New York, pp.481-497.

Standford, G. (1982) Assesment of Soil Nitrogen Availability pp 651-720 In: Stevenson, F.J. (ed.), *Nitrogen in Agricultural Soils.* American Society of Agronomy, Madison, WI, pp.651-720.

Stevenson, F.J. (1986) *Cycles of Soil Carbon, Nitrogen, Phosphorus, Sulfur, Micronutrients.* John Wiley and Sons, New York.

Theodorou, C. and Bowen, G.D. (1983) Nitrogen transformations in first- and second-rotation *Pinus radiata* forest soil. *Australian Forest Research* **13**, 103-12.

Tiedje, J.M. (1982). Denitrification. In: Page, A.L., Miller, R.H. and Keeney, D.R. (eds.), *Methods of Soils Analysis, Part 2. Chemical and Microbiological Properties.* American Society of Agronomy, Madison, Wisconsin, pp.1011-1026.

Tiedje, J.M., Simkins, S. and Groffman, P.M. (1989) *Perspectives of measurement of denitrification in the field including recommended protocols for acetylene based methods.* (In Press).

Vitousek, P.M. and Denslow, J.S. (1986) Nitrogen and phosphorus availability in treefall gaps of a lowland tropical rainforest. *Journal of Ecology* **74**, 1167-1178.

Waring, S.A. and Bremner, J.M. (1964) Ammonium production in soil under waterlogged conditions as an index of nitrogen availability. *Nature* **201**, 951-952.

Westermann, D.T. and Crowthers, S.W. (1980) Measuring soil nitrogen mineralisation under field conditions. *Agronomy Journal* **72**, 1009-1012.

Appendix G TECHNIQUES FOR QUANTIFYING NITROGEN FIXATION

Mark Peoples and Ken Giller

G.1 INTRODUCTION

There is no single "correct" way of measuring biological N_2 fixation. It is unrealistic to expect one technique to provide accurate field measures of N_2 fixation for all the diverse organisms which have the potential to fix N_2 in symbiotic or associative relationships, or in the free-living state. Each methodology has its own unique advantages and limitations which make it more or less suitable for particular species and environments. Appendix G briefly describes several procedures that have been used for measuring symbiotic N_2 fixation by legumes.

Additional information on the application of these techniques may be found in a number of general reviews (eg. Peoples and Herridge, 1990; Giller and Wilson, 1991; Ledgard and Steele, 1992; Hansen, 1994; Herridge and Danso, 1995). Discussion on N_2 fixation by non-legume systems can be found elsewhere (Boddey, 1987).

G.2 NITROGEN DIFFERENCE

A measure of N_2 fixation by a legume can be obtained by growing a companion non-N_2-fixing reference plant in the same soil, under identical conditions (usually in an adjacent plot). The difference in nitrogen yield between the legume and the reference is regarded as the contribution of symbiotic fixation to the legume. It is assumed that the legume and reference crops assimilate the same amount of soil nitrogen (and fertiliser). This relatively simple procedure can be used when only facilities for total nitrogen analysis are available. Measures of N_2 fixation will be most reliable when plants are grown in soils with a poor capacity to supply available nitrogen, where a large proportion of the legume's nitrogen inevitably comes from atmospheric N_2.

Nitrogen difference can provide similar estimates of N_2 fixation to other techniques (eg. Bell *et al.*, 1994); however, the method is prone to underestimate N_2 fixation because legumes often utilise less mineral nitrogen during growth than a non-legume (Herridge *et al.*, 1995). There are also instances in pastures and agroforestry systems where a non-N_2-fixing species can accumulate more nitrogen in their tissues than a legume (Ledgard and Steele, 1992; Ladha *et al.*, 1993). This would result in physiologically impossible "negative" estimates of N_2 fixation using the nitrogen difference procedure.

G.3 ^{15}N-ISOTOPIC TECHNIQUES

The stable isotope of nitrogen, ^{15}N, occurs in the atmospheric N_2 at a constant abundance of 0.3663 atoms %. If the ^{15}N content of plant-available soil nitrogen is significantly different from that of atmospheric N_2, the proportions of legume nitrogen derived from each source can be calculated by comparing the isotopic composition of the nodulated legume with a non-N_2-fixing reference plant totally dependent upon soil nitrogen for growth.

The main advantage of using a ^{15}N-based technique is that measures of N_2 fixation are provided as "time-averaged" estimates that represent the integral of any changes in the N_2-fixing activity that may have occurred during the measurement period. One of the main limitations is the requirement to have ready access to sophisticated and expensive equipment such as a mass spectrometer to accurately quantify the isotopic composition of the legume and reference material.

G.3.1 ^{15}N-enrichment

The use of methods involving artificial adjustment of ^{15}N enrichments to measure N_2 fixation have been reviewed by a number of authors (eg. Chalk, 1985; Danso, 1988; Witty and Giller, 1991). One of the presumed limitations is the high cost of ^{15}N-enriched fertiliser; however, when taken in the context of the overall time invested in planning, and the effort and materials used in establishing a successful field experiment, the monetary value involved in purchasing sufficient ^{15}N for replicate "microplots" is not excessive. A major assumption of the technique is that the legume and the reference absorb the same relative amounts of nitrogen from the added ^{15}N and the soil. The choice of reference plant is the most important single factor affecting the accuracy in estimating N_2 fixation in ^{15}N-enrichment studies (eg. Witty, 1983). Errors due to poorly-matched reference plants can be reduced by applying the ^{15}N-enriched fertiliser in a form which is slowly released (Giller and Witty, 1987), but there are still specific problems when attempting to use the technique with deep-rooted, perennial legumes such as shrubs and trees (Danso *et al.*, 1992).

G.3.2 Natural ^{15}N abundance

The small difference in the natural ^{15}N enrichment of soil mineral nitrogen compared with atmospheric N_2 has been utilised to estimate N_2 fixation (reviewed by Shearer and Kohl, 1986). Although the principles of the technique are the same as ^{15}N-enrichment studies, the main constraints are quite different. The choice of reference appears to be less critical than for ^{15}N-enrichment (Rerkasem *et al.*, 1988; Bergersen *et al.*, 1989; Peoples *et al.*, 1995), but a mass spectrometer capable of precisely measuring differences of 0.00004 atom % ^{15}N is necessary, and care must be taken during sample preparation to avoid isotopic discrimination. Low and variable levels of natural abundance in soil can limit the technique's usefulness (Unkovich *et al.*, 1994); however, this is often not a problem in many agricultural soils (Peoples *et al.*, 1991b). The ^{15}N natural abundance procedure has been demonstrated to provide similar estimates of N_2 fixation to other methodologies (eg. Evans *et al.*, 1987; Bell *et al.*, 1994; Doughton *et al.*, 1995), and the technique has been successfully applied to a number of tropical and temperate crop and forage legumes, and to N_2-fixing trees in both agroforestry systems (Ladha *et al.*, 1993) and natural ecosystems (Hamilton *et al.*, 1993).

G.4 ACETYLENE REDUCTION ASSAY (ARA)

Acetylene is an alternative substrate for the nitrogenase enzyme which reduces acetylene to ethylene. The ARA is a useful diagnostic tool for detecting nitrogenase activity and has been widely used in all areas of N_2 fixation research, but it is now considered to be of limited value for even comparative studies with legumes and it cannot be recommended as a technique for measuring N_2 fixation in the field. Because ARA provides only an instantaneous measure, its accuracy has always been restricted by the need for many repeated determinations to adjust for diurnal, seasonal and age related changes in N_2-fixing activity. But further errors can arise through the use of inappropriate calibration factors relating ethylene production to N_2 fixation, incomplete recovery, disturbance or damage of nodules, and from an acetylene-induced decline in activity during assay (Witty and Minchin, 1988).

G.5 UREIDE METHOD

The xylem sap leaving the roots of a field-grown legume carries nitrogen-compounds to the shoot originating from (i) nodules as assimilation-products of N_2 fixation, and (ii) soil mineral nitrogen taken up by the roots. In some legumes such as soybean, common bean and mung bean, the xylem solutes exported from nodules (the ureides, allantoin and allantoic acid), are very different from the compounds transported from roots (amino acids and nitrate). In these species, the abundance of ureides relative to the other nitrogen components in a xylem sample is indicative of plant dependence upon N_2 fixation (Herridge and Peoples, 1990; Hansen *et al.,* 1993; Herridge *et al.,* 1996), and the technique has been used in breeding programs to select legume genotypes with improved capacities for N_2 fixation (Herridge and Danso, 1995).

Ureide production appears to be restricted to certain taxonomic groups of legumes (Table G.1). Of the papilionoid legumes, all members of the tribes Phaseoleae (except perhaps *Erythrina* spp.) and Desmodieae examined so far transport the products of N_2 fixation as ureides, whilst all members of the Aeschynomeneae, Robinieae, Trifolieae and Vicieae export amides (asparagine and glutamine) from their nodules. Among the Caesalpiniodeae and Mimosoideae there are as yet no confirmed reports of species which transport ureides. Ureides have been reported as minor components of the xylem sap in several species (eg. *Arachis hypogaea, Gliricidia sepium, Sesbania* spp., *Stylosanthes hamata*) where they do not appear to be related to N_2 fixation. In addition, unrelated coloured products have been mistakenly identified as ureides which has probably led to incorrect reports that some species are ureide producers (Peoples *et al.,* 1991a; Herridge *et al.,* 1996).

The ureide method has many advantages over other more complex procedures. Xylem samples can be recovered from field-grown legumes by applying a mild vacuum (see G.5.1), the three nitrogen components of xylem sap (ureides, amino acids and nitrate) can be measured separately by simple colorimetric assays, and plant reliance upon N_2 fixation for growth (Pfix) can calculated by relating the determined xylem sap composition to glasshouse prepared calibration curves. The relative ureide composition of xylem sap is not affected by the time of sampling during the day (Peoples and Herridge, 1990). The major disadvantage is that the procedure provides only short-term estimates of Pfix. If seasonal determinations are required rather than comparative measures of treatment effects, repeated xylem collections must be taken (Herridge *et al.,* 1990).

G.5.1 Sampling of N-solutes

It is possible to collect xylem exudate as in bleeds spontaneously from intact root stumps of crop legumes following decapitation of the shoot from plants growing in the field in the humid tropics. In other harsher environments the xylem contents can be recovered from detached stems or branch segments from woody legumes by mild vacuum. There appears to be only slight differences between the N-solute compositions of root-bleeding exudates and vacuum-extracted saps. Usually 4 replicates of 4 - 10 plants will be used for sap collections. The plant material can be retained to provide a measure of crop N. Only 1 or 2 sap collections may be required (e.g. during flowering or early seed-filling) to compare the effect of treatments (e.g. inoculation or N-fertiliser) on N_2 fixation, but repeated sampling (4 - 8 times) during plant growth will be required to determine seasonal inputs of fixed N.

Procedure for root-bleeding exudate

1. Cut the shoot close to ground level with a sharp blade or pair of secateurs.

2. Place a silicon or latex rubber tubing sleeve with an internal diameter slightly smaller than the stem over the exposed root stump.

3. Collect the sap exuded into the tubing after 10-20 min using a Pasteur pipette or syringe.

Procedure for vacuum-extracted exudate

1. Detach a stem or branch segment (> 3 mm diameter).

2. Immediately insert the detached stem into a close fitting rubber sleeve connected to a syringe needle (19 or 20 gauge).

3. Push the needle through the rubber stopper of a 5 ml Vacutainer (commonly used for collecting blood samples; Becton Dickinson, Rusherford, USA) which has been linked to a vacuum pump via another syringe needle connection and a flexible plastic-tubing line. The source of vacuum may be a hand-held vacuum pump (Nalgene, Sybron Corp, USA), or consist of a laboratory vacuum pump powered by a portable petrol-run generator or car battery if an electricity supply is not available.

4. Apply partial vacuum and successively cut short segments (3 - 4 cm) of stem from the top to the bottom of the shoot. Xylem sap will be collected within the Vacutainer.

Note: Sap collections should be restricted to between 9 am and 4 pm. Vacuum extraction must occur within 5 - 10 min of stem detachment.

G.5.2 Sap storage

The N-solute composition of xylem sap is stable for only a few hours at temperatures above 25°C. Sap samples should be placed in an ice-box immediately after collection and remain

Table G.1 Tropical legumes which transport the products of N_2-fixation mainly as ureides or amides (Giller and Wilson, 1991).

Ureide producers	Amide producers
Grain legumes	
Phaseoleae: *Cajanus cajan,* *Glycine max,* *Lablab purpureus,* *Macrotyloma geocarpa, M. uniflorum,* *Phaseolus lunatus, P. vulgaris,* *Psophocarpus tetragonolobus,* *Vigna aconitifolia, V. angularis, V. mungo, V. radiata, V. subterranea, V. trilobata, V. umbellata, V. unguiculata* Indigofereae: *Cyamopsis tetragonoloba*	Aeschynomeneae: *Arachis hypogaea* Cicereae: *Cicer arietinum* Genisteae: *Lupinus mutabilis* Vicieae: *Lathyrus sativus,* *Lens culinaris, L. esculenta,* *Vicia faba*
Forage legumes	
Desmodieae: *Desmodium discolor, D. uncinatum* Phaseoleae: *Calopogonium caeruleum* *Centrosema pubescens,* *Macroptilium atropurpureum,* *Pueraria javanica, P. phaseoloides*	Aeshynomeneae: *Arachis glabrata, A. pintoi* *Zornia spp.* Crotalarieae: *Crotalaria spp.* Trifolieae: *Trifolium pratense, T. repens, T. subterraneum*
Shrub and tree legumes	
Desmodieae: *Codariocalyx gyroides,* Desmodium resonsinii	Aeschynomeneae: *Aeschynomene indica* Acacieae: *Acacia alata, A. auriculiformis, A. extensa, A. pulchella, A. unisauvis* Ingeae: *Calliandra calothyrsus* Mimoseae: *Leucaena diversifolia, L. leucocephala, L. macrophylla, Prosopis juliflora* Robinieae: *Gliricidia sepium, Sesbania grandiflora, S. sesban, S. rostrata*

chilled until placed in a freezer for long-term storage. If this is not possible, samples may be stabilised by adding an equal volume of absolute ethanol.

G.5.3 Analyses

Apart from the specific chemicals and acids required for the three assays, the following items will be required before commencing N-solute analyses:

- A weighing balance, preferably accurate to 0.1 mg
- Test-tubes (e.g. 150 x 18 mm), racks, and a boiling-water bath
- Micropipettes and/or dispensers to cover the range 0.01 - 0.2 ml and 0.5 - 5 ml
- Vortex mixer and cold-water ice bath
- Spectrophotometer or colorimeter.

A volume of > 0.2 ml of sap will be required for the three analyses (0.02 - 0.5 ml of sap is used for each assay). Usually 0.5 - 1 ml might be expected to be collected from mature field-grown soybean or pigeonpea, but it is often more difficult to collect sap from other species such as cowpea and mung bean and larger numbers of plants may be required to obtain sufficient sap for analysis.

(a) Total ureides are measured by the method of Young and Conway (1942)

(b) Amino acids are determined with ninhydrin (Yemm and Cocking, 1955) using a 50:50, asparagine:glutamine standard

(c) Nitrate is measured with salicylic acid (Cataldo *et al.*, 1975).

Full details of the assays and the use of the ureide method are described in Peoples *et al.* (1989), copies of which may be obtained by writing to ACIAR (GPO Box 1571, Canberra, ACT 2601, Australia).

Calculations

The ureide content of a sample is expressed in terms of a proportional value of ratio whereby the amount of N present as ureide (4 atoms per molecule) is related to total sap N. Calculation of a relative ureide index (RUI) from the individual solute concentrations ([ ]) in moles is determined as:

RUI (%) = {100 (4 x [ureide])} / {(4 x [ureide]) + [amino acid] + [nitrate]}

Experiments calibrating relative ureide index against ^{15}N-derived estimates of N_2 fixation have been undertaken for several legume species. However, it appears that many ureide-exporting species have similar relationships to soybean (Peoples and Herridge, 1990), therefore only calibration functions prepared for soybean (Herridge and Peoples, 1990) will be presented below.

In soybean there were discernible effects of plant age on N-solute compositions resulting in two sets of calibrations - one for vegetative and flowering plants, and another during pod-fill.

For each sap collection method:

Root-bleeding exudate

The functions describing the relationship between Pfix (the proportion of plant N derived from N_2 fixation) and the relative abundance of ureides (x) are:

Pfix (%) = 1.2 (x - 5) for vegetative and flowering stages
Pfix (%) = 1.5 (x - 21) for plants during pod-fill

Vacuum-extracted exudate

Pfix (%) = 1.6 (x - 8) for vegetative and flowering stages
Pfix (%) = 1.6 (x - 16) for plants during pod-fill

Measurements of N_2 fixation by field-grown legumes using such glasshouse-derived calibrations have been shown to be very similar to estimates calculated from sophisticated ^{15}N techniques (e.g. Herridge *et al.*, 1990).

References

Bell, M.J., Wright, G.C., Suryantini and Peoples, M.B. (1994) The N_2-fixing capacity of peanut cultivars with differing assimilate partitioning characteristics. *Australian Journal of Agricultural Research* **45**, 1455-1468.
Bergersen, F.J., Brockwell, J., Gault, R.R., Morthorpe, L., Peoples, M.B. and Turner, G.L. (1989) Effects of available soil nitrogen and rates of inoculation on nitrogen fixation by irrigated soybeans and evaluation of $\delta^{15}N$ methods for measurement. *Australian Journal of Agricultural Research* **40**, 763-780.
Boddey, R.M. (1987) Methods for quantification of nitrogen fixation associated with Gramineae. *CRC Critical Reviews in Plant Sciences* **6**, 209-266.
Cataldo, D.A., Haroon, M., Schrader, L.E. and Youngs, V.L. (1975) Rapid colorimetric determinations of nitrate in plant tissue by nitration of salicylic acid. *Communications of Soil Science and Plant Analysis* **6**, 71-80.
Chalk, P.M. (1985) Review. Estimation of N_2 fixation by isotope dilution: An appraisal of techniques involving ^{15}N enrichment and their application. *Soil Biology and Biochemistry* **17**, 389-410.
Danso, S.K.A. (1988) The use of ^{15}N enriched fertilizers for estimating nitrogen fixation in grain and pasture legumes. In: Beck, D.P. and Materon, L.A. (eds.), *Nitrogen Fixation by Legumes in Mediterranean Agriculture*. Martinus Nijhoff, Dordrecht, Netherlands, pp.345-458.
Danso, S.K.A., Bowen, G.D. and Sanginga, N. (1992) Biological nitrogen fixation in trees in agro-ecosystems. *Plant and Soil* **141**, 177-196.
Doughton, J.A., Saffigna, P.G., Vallis, I. and Mayer, R.J. (1995) Nitrogen fixation in chickpea. II. Comparisons of ^{15}N enrichment and ^{15}N natural abundance methods for estimating nitrogen fixation. *Australian Journal of Agricultural Research* **46**, 225-236.

Evans, J., Turner, G.L., O'Connor, G.E. and Bergersen, F.J. (1987) Nitrogen fixation and accretion of soil nitrogen by field-grown lupins (*Lupins angustifolius*). *Field Crops Research* **16**, 309-322.

Giller, K.E. and Witty, J.F. (1987) Immobilized ^{15}N-fertiliser sources improve the accuracy of field estimates of N_2-fixation by isotope dilution. *Soil Biology and Biochemistry* **19**, 459-463.

Giller, K.E. and Wilson, K.J. (1991) *Nitrogen Fixation in Tropical Cropping Systems*. CAB International, Wallingford.

Hamilton, S.D., Hopmans, P., Chalk, P.M. and Smith, C.J. (1993) Field estimation of N_2 fixation by *Acacia* spp. using ^{15}N isotope dilution and labelling with ^{35}S. *Forest Ecology and Management* **56**, 297-313.

Hansen, A.P. (1994) *Symbiotic N_2 Fixation of Crop Legumes*. Hohenheim Tropical Agricultural Series. Margraf Verlag, Weikersheim, Germany, 248p.

Hansen, A.P., Rerkasem, B., Lordkaew, S. and Martin, P. (1993) Xylem-solute technique to measure N_2 fixation by *Phaseolus vulgaris* L.: calibration and sources of error. *Plant and Soil* **150**, 223-231.

Herridge, D.F. and Peoples, M.B. (1990) Ureide assay for measuring nitrogen fixation by nodulated soybean calibrated by ^{15}N methods. *Plant Physiology* **93**, 495-503.

Herridge, D.F., Bergersen, F.J. and Peoples, M.B. (1990) Measurement of nitrogen fixation by soybean in the field using the ureide and natural ^{15}N abundance methods. *Plant Physiology* **93**, 708-716.

Herridge, D.F. and Danso, S.K.A. (1995) Enhancing crop legume N_2 fixation through selection and breeding. *Plant and Soil* **174,** 51-82.

Herridge, D.F., Marcellos, H., Felton, W.L., Turner, G.L. and Peoples, M.B. (1995) Chickpea increases soil-N fertility in cereal systems through nitrate sparing and N_2 fixation. *Soil Biology and Biochemistry* **27**, 545-551.

Herridge, D.F., Palmer, B., Nurhayati, D.P. and Peoples, M.B. (1996) Evaluation of the xylem ureide method for measuring N_2 fixation in six tree legume species. *Soil Biology and Biochemistry* (in press).

Ladha, J.K., Peoples, M.B., Garrity, D.P., Capuno, V.T. and Dart, P.J. (1993) Estimating dinitrogen fixation of hedgerow vegetation in an alley-crop system using the ^{15}N natural abundance method. *Soil Science Society of America Journal* **57**, 732-737.

Ledgard, S.F. and Steele, K.W. (1992) Biological nitrogen fixation in mixed legume/grass pastures. *Plant and Soil* **141**, 137-153.

Peoples, M.B. and Herridge, D.F. (1990) Nitrogen fixation by legumes in tropical and subtropical agriculture. *Advances in Agronomy* **44**, 155-223.

Peoples, M.B., Faizah, A.W., Rerkasem, B. and Herridge, D.F. (1989a) *Methods for Evaluating Nitrogen Fixation by Nodulated Legumes in the Field.* ACIAR Monograph No 11, ACIAR, Canberra. 76pp.

Peoples, M.B., Atkins, C.A., Pate, J.S., Chong, K., Faizah, A.W., Suratmini, P., Nurhayati, D.P., Bagnall, D.J. and Bergersen, F.J. (1991*a*) Re-evaluation of the role of ureides in the xylem transport of nitrogen in *Arachis* species. *Physiologia Plantarum* **83**, 560-567.

Peoples, M.B., Bergersen, F.J., Turner, G.L., Sampet, C., Rerkasem, B., Bhromsiri, A., Nurhayati, D.P., Faizah, A.W., Sudin, M.N., Norhayati, M. and Herridge, D.F. (1991*b*) Use of natural enrichment of ^{15}N in plant available soil N for the measurement of symbiotic N_2 fixation. In: *Stable Isotopes in Plant Nutrition, Soil Fertility and Environmental Studies.*, pp. 117-130, IAEA, Vienna.

Peoples, M.B., Bergersen, F.J., Brockwell, J., Fillery, I.R.P. and Herridge, D.F. (1995) Management of nitrogen for sustainable agricultural systems. In *Nuclear Methods in Soil-Plant Aspects of Sustainable Agriculture* , IAEA-TECDOC-785, FAO/IAEA, Vienna, Austria, pp. 17-35.

Rerkasem, B., Rerkasem, K., Peoples, M.B., Herridge, D.F. and Bergersen, F.J. (1988) Measurement of N_2 fixation in maize (*Zea mays* L.) - ricebean (*Vigna umbellata* [Thumb] Ohwi Ohashi) intercrops. *Plant and Soil* **108**, 125-135.

Shearer, G. and Kohl, D.H. (1986) Review. N_2-fixation in field settings: Estimations based on natural ^{15}N abundance. *Australian Journal of Plant Physiology* **13**, 699-756.

Unkovich, M.J., Pate, J.S., Sanford, P. and Armstrong, E.L. (1994) Potential precision of the $\delta^{15}N$ natural abundance method in field estimates of nitrogen fixation by crop and pasture legumes in south-west Australia. *Australian Journal of Agricultural Research* **45**, 119-132.

Witty, J.F. (1983) Estimating N_2-fixation in the field using ^{15}N-labelled fertiliser: some problems and solutions. *Soil Biology and Biochemistry* **15**, 631-639.

Witty, J.F. and Minchin, F.R. (1988) Measurement of nitrogen fixation by the acetylene reduction assay; myths and mysteries. In: Beck, D.P. and Materon, L.A. (eds.), *Nitrogen Fixation by Legumes in Mediterranean Agriculture*. Martinus Nijhoff, Dordrecht, Netherlands, pp.331-344.

Witty, J.F. and Giller, K.E. (1991) Evaluation of errors in the measurement of biological nitrogen fixation using ^{15}N fertiliser. In: *The Use of Stable Isotopes in Plant Nutrition, Soil Fertility and Environmental Studies.*, pp. 59-72, IAEA, Vienna.

Yemm, E.W. and Cocking, E.F. (1955) The determination of amino acids with ninhydrin. *Analyst* **80**, 209-213.

Young, E.G. and Conway, C.F. (1942) On the estimation of allantoin by the rimini-schryver reaction. *Journal of Biological Chemistry* **142**, 839-853.

Appendix H MOST PROBABLE NUMBER COUNTS OF RHIZOBIA IN SOILS

Paul Woomer

H.1 BACKGROUND

An estimation of the total viable rhizobia in soils and other test substrates can be obtained through application of the most-probable-number technique (MPN). This technique is used to estimate microbial population sizes when direct quantitative assessment of individual cells is not possible.

A suitable legume host is first cultured on rhizobia-free media, then a test substrate is serially diluted and applied to the root systems of the legume hosts. After 21 days the pattern of presence and absence of root nodules is recorded, and these data used to derive a population estimate of the original test substrate.

The MPN technique is based on the mathematical approaches of Halvorson and Zeigler (1933). Cochran (1950) later identified estimation of error and mean separation procedures. Computer programs are now available which combine these operations resulting in greater flexibility of experimental designs and scientific procedures (Woomer *et al.*, 1990). Tests of experimental technique that allow researchers to either accept or reject experimental results are also available (Halvorson and Moeglein, 1940; Woodward, 1957; Stevens, 1958; deMan, 1975; Scott and Porter, 1986; Woomer *et al.*, 1988a).

The need for the application of rhizobial inoculants and the magnitude of the response to applied rhizobia are determined by the species diversity and population sizes of indigenous soil rhizobia and the availability of mineral nitrogen within the soil system (Theis *et al.*, 1991). The legume/rhizobia symbiosis is characterised by specificity between the host and microsymbiont; these are identified as cross inoculation groups (FAO, 1984). The selection of a range of legumes characteristic of individual cross-inoculation groups as hosts for MPN assays allows for the species composition and population sizes to be characterised (Woomer *et al.*, 1988b). The lack of rhizobia within a soil allows for the comparison of inoculated and uninoculated treatments as a direct measure of biological nitrogen fixation using the nitrogen difference method (see Appendix G).

H.2 EXPERIMENTATION

H.2.1 Experimental design

The design of a MPN determination involves the selection of the base dilution ratio (A); the selection of the number of units at each dilution level (N); the determination of the need of an initial dilution and its value (I) and the selection of the inoculation volume (V). In this procedure, a five-fold dilution series is inoculated onto the root system of 4 replicate plants per dilution level with 1 ml per experimental test unit.

H.2.2 Materials

The following materials are required to perform the MPN determination illustrated in Figure H.1.

1. Stock solution: 0.125 g K_2HPO_4 and 0.05 g $MgSO_4.7H_2O$ in 1000 ml of distilled water in a sterile 1 litre wide neck Erlenmeyer flask.
2. 100 g (dw) of fresh soil and 400 ml stock solution in a sterile 1 litre wide neck Erlenmeyer flask to initiate the 1:5 dilution series.
3. 5 sterile 100 ml screw top Erlenmeyer flasks and stoppers each containing 20 ml of sterile mineral nutrient diluent.
4. 1 sterile 5 ml wide-mouth pipette.
5. 5 sterile 5 ml pipettes.
6. 1 sterile 1 ml pipette.
7. 1 growth pouch rack (Somesegaran and Hoben, 1985).
8. 28 growth pouches containing uniform nodulating legume seedlings.
9. 56 surface sterilised, pre-germinated legume seeds.

H.2.3 Plant preparation

In MPN plant-infection procedures, nodulation of the legume host serves as a selective media in determining the presence or absence of the rhizobia at a given dilution level. Therefore, the choice of a specific legume host greatly influences the experimental results. More information on the specificity of legume hosts and the selection of these hosts as an indicator of rhizobial species may be obtained from FAO (1984) and Woomer *et al.* (1988a). In this procedure, the host plants are cultured in "growth pouches" (Weaver and Frederick, 1972; Somesegaran and Hoben, 1985) available from Vaughan's Seed Company, 5300 Katrine Ave., Downers Grove, IL, USA (approximate cost: $30/100). A technique for the culture of small-seeded legumes in glass tubes is available in Brockwell (1963) and Brockwell *et al.* (1975).

1 Surface sterilise seeds by emersion in a 1.5% solution of sodium hypochlorite (or an equivelent concentration of household bleach) for 4 minutes followed by 5 rinses with sterile distilled water. For "hard seeded" legumes, emersion in concentrated sulphuric acid for 4-25 minutes followed by 8 rinses may be substituted as a means of simultaneously surface sterilising and scarifying the seeds.

2 Transfer the seeds to a sterile germination vessel.

3 Upon emergence of the radical, transfer 2 germinated seeds to the trough of growth pouch containing 75-100 ml of sterile nitrogen-free nutrient solution (Woomer *et al.*, 1988) or rhizobia-free water, taking care that the radical is in partial contact with the paper wick of the growth pouch. Do not allow the growth pouch wick to dry, additional irrigation is required every 4-10 days. Culture host plants for 1 week prior to inoculation in a clean glasshouse or growth room.

H.2.4 Soil sampling, preparation and storage prior to dilution

When collecting soil samples, care must be taken to collect soils with soil sampling equipment that is not contaminated. It is not necessary to clean equipment between subsamples but all sampling equipment should be sprayed with alcohol and flame sterilised between different sites or treatments. Soils should not be air dried prior to dilution, and are best stored under refrigeration. Dilute and inoculate the soil samples within 48 hours of collection. Prior to dilution, remove stones, gravel, large plant roots, root nodules and non-decomposed organic residues by hand. Crush larger soil aggregates but it is not necessary to sieve the soil since the soil disperses during the first dilution step.

H.2.5 Preparation of the dilution series, root inoculation and observations of nodulation

Procedure

1 Place 100 g (dw) soil into a large mouthed jar or flask containing 400 ml of sterile diluent. Low concentrations of mineral salts are used as diluents instead of distilled water in order to avoid differences in osmotic potential across the dilution series.

2 Mix the soil and diluent using a wrist action shaker set at about 100 Hz for 20-25 min. This is the first dilution step.

3 Transfer 5 ml of the first dilution step to a 100 ml sterile screw top Erlenmeyer flask containing the 20 ml of diluent using a sterile, wide mouthed pipette. The soil must not settle to the bottom of the vessel before the aliquot is withdrawn. Close the test tube and agitate using a vortex mixer, or by wrist action. This is the second dilution step.

4 Continue the dilution series to 5^{-6} (see Figure H.1) using a sterile, narrow mouthed pipette for each additional dilution step. Again, care must be taken to avoid sedimentation prior to the withdrawal of the transferred aliquots. These are the third through sixth dilution steps.

5 Apply 1.0 ml of each dilution step directly to the legume root systems within the growth pouches using a sterile 1.0 ml pipette. A single pipette can be used for the entire dilution series of each test substrate if test units are inoculated in order from the highest (5^{-6}) to the lowest (5^{-1}) dilution, and the pipette rinsed twice prior to the transfer of the 1.0 ml inoculant to the culture units by drawing and releasing the dilution back into the vessel into and from the pipette before inoculation at each new dilution level.

6 Place growth pouches in a clean growth chamber or glasshouse. Check for the appearance of root nodules daily. In general, plant infection counts yield results between 14 and 21 days.

7 Record data by regularly examining the root systems for the presence of root nodules. Culture units that yield negative results should not be immediately discarded, but should continue to be checked daily until 28 days after inoculation to insure that no delayed nodule initiation takes place.

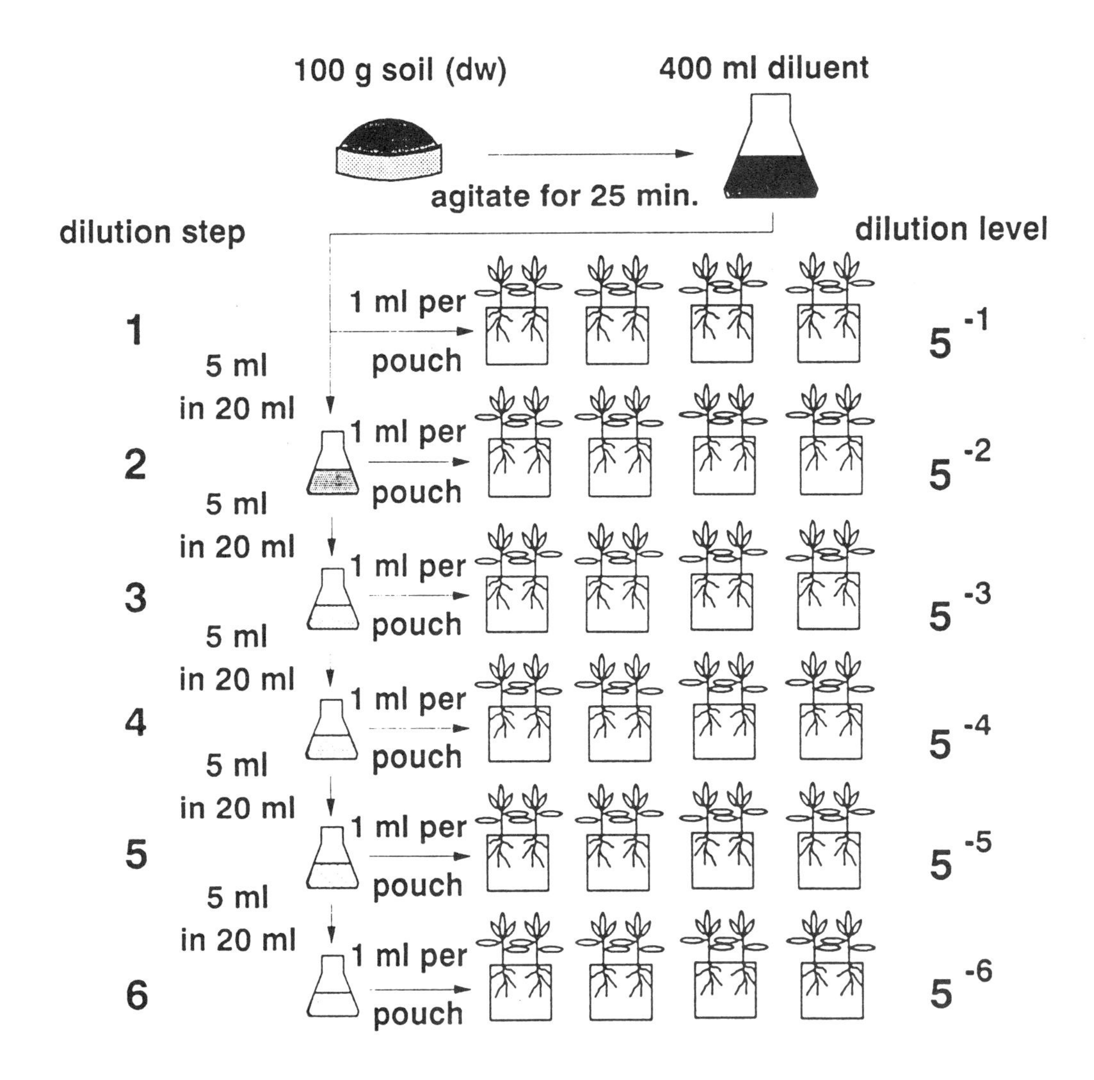

Figure H.1 Diagrammatic procedure of an most-probable-number plant-infection count with 6 fivefold dilution steps and 4 units per dilution level.

H.3 ASSIGNING TABULAR POPULATION ESTIMATES TO RESULTS

The population estimates are assigned by locating the experimental results on Table H.1. When the correct code is located, the researcher obtains a population estimate of the original test substrate directly from the table. In order to develop a population estimate for a fivefold serial dilution with 4 replicates and experimental results of 4-4-3-2-0-0, the operator searches

Table H.1 The most-probable-number and confidence intervals for a six step, fivefold dilution series with 4 test units per dilution level each inoculated with 1.0 ml.

Positive at dilution 1 2 3 4 5 6	MPN estimate (cells/g)	Confidence interval (p = 0.05)
0 1 0 0 0 0	1.0	0.4-2.9
1 0 0 0 0 0	1.1	0.4-3.2
1 0 1 0 0 0	2.3	0.8-6.5
1 1 0 0 0 0	2.3	0.8-6.6
2 0 0 0 0 0	2.6	0.9-7.4
2 0 1 0 0 0	3.9	1.4-11.2
1 2 0 0 0 0	3.5	1.2-10.2
2 1 0 0 0 0	4.0	1.4-11.5
2 1 1 0 0 0	5.4	1.9-15.6
2 2 0 0 0 0	5.5	1.9-16.0
3 0 0 0 0 0	4.6	1.6-13.2
3 0 1 0 0 0	6.3	2.2-18.1
3 1 0 0 0 0	6.5	2.3-19.0
3 1 1 0 0 0	8.4	2.9-24.3
3 2 0 0 0 0	8.7	3.0-25.1
3 2 1 0 0 0	10.9	3.8-31.3
3 3 0 0 0 0	11.2	3.9-32.6
4 0 0 0 0 0	8.0	2.8-23.3
4 0 1 0 0 0	10.8	3.8-31.2
4 1 0 0 0 0	11.4	4.0-33.0
4 1 1 0 0 0	15.1	5.2-43.5
4 2 0 0 0 0	16.2	5.6-46.8
4 1 2 0 0 0	19.6	6.8-56.5
4 2 1 0 0 0	21.5	7.5-62.1
4 2 1 1 0 0	27.7	9.6-79.8
4 2 2 0 0 0	28.3	9.8-81.5
4 3 0 0 0 0	24.2	8.4-69.7
4 3 0 1 0 0	31.9	11.1-92.0
4 3 1 0 0 0	32.8	11.4-94.6
4 3 1 1 0 0	42.2	14.6-121.5
4 3 2 0 0 0	43.6	15.1-125.6
4 3 2 1 0 0	54.4	18.9-156.8
4 3 3 0 0 0	56.5	19.6-163.0
4 4 0 0 0 0	40.4	14.0-116.5
4 4 0 1 0 0	54.2	18.8-156.3
4 4 1 0 0 0	57.1	19.8-164.7
4 4 1 0 1 0	74.6	25.9-215.0
4 4 1 1 0 0	75.5	26.2-217.6
4 4 2 0 0 0	81.3	28.2-234.3
4 4 2 0 1 0	106.0	36.8-305.5
4 4 1 2 0 0	98.2	34.0-283.1
4 4 2 1 0 0	107.9	37.4-311.0
4 4 2 1 1 0	138.6	48.1-399.7
4 4 2 2 0 0	141.7	49.1-408.5
4 4 3 0 0 0	121.1	42.0-349.1
4 4 3 0 1 0	159.8	55.4-460.8
4 4 3 1 0 0	164.7	57.1-474.9
4 4 3 1 1 0	211.4	73.3-609.5
4 4 3 2 0 0	218.4	75.8-629.7
4 4 3 2 1 0	273.0	94.7-787.2
4 4 3 3 0 0	283.4	98.3-817.1
4 4 4 0 0 0	202.6	70.3-584.0
4 4 4 0 1 0	272.4	94.5-785.4
4 4 4 1 0 0	287.1	99.6-287.6
4 4 4 1 0 1	375.0	130.1-1081.0
4 4 4 1 1 0	379.9	131.8-1095.1
4 4 4 2 0 0	409.1	141.9-1179.5
4 4 4 2 0 1	534.9	185.5-1541.9
4 4 4 1 2 0	495.8	172.0-1429.3
4 4 4 2 1 0	544.6	188.9-1570.1
4 4 4 2 1 1	698.4	242.3-2013.5
4 4 4 2 2 0	718.0	249.0-2069.8
4 4 4 3 0 0	610.5	211.8-1760.1
4 4 4 3 0 1	810.7	281.2-2337.2
4 4 4 3 1 0	830.2	1074.4-2393.5
4 4 4 3 1 1	1074.4	372.7-3097.2
4 4 4 3 2 0	1113.4	386.2-3209.8
4 4 4 3 2 1	1386.8	481.1-3998.0
4 4 4 3 3 0	1445.4	501.4-4166.9
4 4 4 4 0 0	1035.3	359.1-2984.6
4 4 4 4 0 1	1386.8	481.1-3998.0
4 4 4 4 1 0	1464.9	508.2-4223.2
4 4 4 4 1 1	1992.2	691.1-5743.4
4 4 4 4 2 0	2109.4	731.7-6081.1
4 4 4 4 1 2	2578.2	894.3-7432.6
4 4 4 4 2 1	2812.6	975.6-8108.4
4 4 4 4 2 2	3750.1	1300.8-10810.9
4 4 4 4 3 0	3281.3	1138.2-9459.7
4 4 4 4 3 1	4375.0	1517.6-12612.6
4 4 4 4 3 2	5937.6	2059.6-17117.3
4 4 4 4 3 3	7500.1	2601.6-21621.7
4 4 4 4 4 0	5937.6	2059.6-17117.3
4 4 4 4 4 1	8750.0	3035.2-25225.2
4 4 4 4 4 2	16250.1	5636.8-46846.7
4 4 4 4 4 3	20000.0	6937.5-57657.3

the first column for those experimental results and then obtains the population estimate from the corresponding adjacent column (218 cells/g soil).

The population estimates in Table H.1 assume that the plant roots of each experimental unit were inoculated with 1.0 ml of the diluted test substrate and that there was no pre-dilution of the test substrate prior to the serial dilution. If an inoculant volume other than 1.0 ml was applied to the culture units, the population estimate must be divided by the inoculation volume. For example, if 2.5 ml was applied, the tabular population estimate is divided by 2.5. If an pre-dilution was prepared the tabular population estimate is multiplied by that dilution value. For example, if a test substrate is diluted 100-fold prior to the fivefold dilution series and the pre-dilution was used as the source of the lowest dilution, the tabular value obtained is multiplied by 100.

H.3.1 Constructing confidence limits

The confidence limits are constructed by dividing and multiplying the population estimate by a confidence factor (Cochran, 1950). The confidence limits (p = 0.05) of the MPN estimates are given in Table H.1. For the fivefold dilution series with 4 replicates, the confidence factor is 2.88 (p=0.05). The experimental results 4-4-3-2-0-0 yield a tabular MPN estimate of 218, the lower confidence limit (p=0.05) equals 218 / 2.88, or 76. The upper limit is 218 x 2.88 or 627. These results may be expressed as 218 (76-627, p=0.05).

References

Brockwell, J. (1963) Accuracy of a plant-infection technique for counting populations of *Rhizobium trifolii*. *Applied Microbiology* **11**, 377-383.

Brockwell, J., Diatloff, A. Garcia, A.L. and Robinson, A.C. (1975) Use of wild soybean (*Glycine ussuriensis* Regel and Maack) as a test plant in dilution-nodulation frequency tests for counting *Rhizobium japonicum*. *Soil Biology and Biochemistry* **7**, 305-311.

Cochran, W.G. (1950) Estimation of bacterial densities by means of the "most probable number". *Biometrics* **6**, 105-116.

deMan, J.C. (1975) The probability of most probable numbers. *European Journal of Applied Microbiology* **1**, 67-78.

FAO (Food and Agriculture Organisation of the United Nations) (1984) *Legume Inoculants and Their Use*. FAO, Rome.

Halvorson H.O and Moeglein, A. (1940) Application of statistics to problems in bacteriology. V. The probability of occurrence of various experimental results. *Growth* **4**, 157-168.

Halvorson, H.O. and Ziegler, N.R. (1933) Applications of statistics to problems in bacteriology. I. A means of determining bacterial population by the dilution method. *Journal of Bacteriology* **25**, 101-121.

Somesegaran P. and Hoben, H.J. (1985) *Methods in Legume-Rhizobium Technology*. University of Hawaii NifTAL Project and MIRCEN. Paia, Hawaii, USA.

Theis, J.E., Singleton, P.W. and Bohlool, B.B. (1991) Influence of the size of indigenous rhizobial populations on the establishment and symbiotic performance of introduced rhizobia on field-grown legumes. *Applied and Environmental Microbiology* **57**, 19-28.

Weaver, R.W. and Frederick, L.R. (1972) A new technique for most-probable-number counts of rhizobia. *Plant and Soil* **36**, 219-222.

Woomer, P., Singleton, P.W. and Bohlool, B.B. (1988a) Reliability of the most-probable-number technique for enumerating rhizobia in tropical soils. *Applied and Environmental Microbiology* **54**, 1494-1497

Woomer, P, Singleton, P.W. and Bohlool, B.B. (1988b) Ecological indicators of native rhizobia in tropical soils. *Applied and Environmental Microbiology* **54**, 1112-1116.

Woomer, P, Bennett, J. and Yost, R. (1990) Overcoming the inflexibility of Most-Probable-Number procedures. *Agronomy Journal* **82**, 349-353.

Appendix I THE STUDY OF PHOSPHORUS CYCLES IN ECOSYSTEMS

Holm Tiessen

I.1 INTRODUCTION

Phosphorus (P) is one of the main limiting elements for crop production. In many natural ecosystems P availability limits overall ecosystem productivity through its effect on plant production and nitrogen fixation. Plants take up P from the soil solution. While solution P constitutes only a minute fraction of total soil P, it is constantly replenished by the hydrolysis of labile inorganic P (Pi) or by mineralisation of organic P (Po). These fractions, in turn, are in exchange with more stable P compounds through slow reactions. Therefore, the P supply to vegetation depends on the quantities of labile Pi, the rates of transformations between labile and more stable Pi, and the size and rates of transformation of the mineralisable Po pool in the soil.

I.2 SOIL TESTS FOR INORGANIC P AVAILABILITY

Soil tests for plant available Pi need to include solution Pi and varying amounts of Pi associated with the soils solid phase, and they must be simple enough for routine application. This is achieved with single extractants designed to moderately lower or raise pH of the soil solution, introduce anions that solubilise P by competing with P sorption sites (HCO_3, SO_4), or by chelating (EDTA) cations or lowering the solubility of cations (F^-) that bind P in the soil. Examples are the acid ammonium fluoride (Bray and Kurtz, 1945), bicarbonate (Olsen *et al.*, 1954; Dabin, 1967), lactate (Egner *et al.*, 1960), or chelating (Onken *et al.*, 1980) extracts. An alternative extractant, ion exchange resins, simply constitute a sink for solution Pi and thereby drive soluble or "exchangeable" P into solution (Amer *et al.*, 1955). To some extent, the extraction of P by resin is affected by the saturating anion on the resin, and by the presence of different cations in the soil solution (Curtin *et al.*, 1987). Bicarbonate resin is frequently used in temperate soils but a neutral anion such as Cl^- is to be preferred in acid soils with low buffering capacity to avoid uncontrolled and variable pH rises from the released HCO_3.

Although some extractants have been chosen to simulate the action of plant roots which produce bicarbonate and various organic acids, in reality, the processes occurring in the vicinity of plant roots are likely to be much more complex, involving microbes, mycorrhiza, specific root exudates and pH control (López-Hernández and Flores-Aguilar, 1979). The biennial pigeonpea (*Cajanus cajan*) may have special abilities to sequester Fe-P, that is unavailable to other plants, by excretion of specialised chelating substances (Ae *et al.*, 1990).

Chemical extractants clearly cannot mimic such complex processes. The success of simple P tests in giving 50-60% accurate predictions for fertiliser requirements in many moderately weathered soils depends on the extraction of a quantity of soil P that is somehow related to that portion of soil P which is plant available. The regression between these two quantities is established over years of agronomic experimentation and testing of fertiliser responses for specific crops on defined soil types with appropriately matched extractants (Sharpley *et al.*, 1985). On weakly weathered Ca-rich soils, acid extractants (Bray and Kurtz, 1945; Nelson *et al.*, 1953) are mismatched because they can dissolve P that is not plant available, and that would not be extracted by more alkaline solutions (Olsen *et al.*, 1954). In acid soils, correlations between mild acid and alkaline extractants may be closer (Sharpley *et al.*, 1985), but in acid tropical soils that show rapid P transformations and substantial P sorption, extraction methods for P fertility assessment have often been unsuccessful, and fertiliser responses have been erratic (Ayodele and Agboola, 1983).

Under tropical conditions, rapid mineralisation of organic matter from slashed vegetation or crop residues can liberate sufficient P for plant growth to obscure the role of inorganic P extracted by P tests (Adepetu and Corey, 1976; Agboola and Oko, 1976). In addition to the faulty assessment of available P, fertiliser responses may be altered by strong sorption of P. In order to correct for such P "fixation", the fractional recovery of added fertiliser by a P test (Agboola and Ayodele, 1983), the amount of P required to raise solution P (Van der Zaag *et al.*, 1979; Juo and Fox, 1977) or extractable P (Agboola and Ayodele, 1983) to some chosen value, direct measures of P sorption (Roche, 1983; Fox and Kamprath, 1970), or determination of P-sorbing Fe (Anyaduba and Adepetu, 1983; LeMare, 1981) or Al (Roche *et al.*, 1983) can be combined with P tests.

I.3 PHOSPHORUS "FIXATION"

Phosphorus "fixation" is, as any sorption process, reversible, and in many weathered soils, Fe- and Al-phosphates are also a potential source of plant available P (Hughes, 1982; Hughes and LeMare, 1982). LeMare (1982) indicated that particularly organic matter - associated Fe and Al may be involved in the reversible sorption of P. One of the problems in assessing the reversibility of sorption is that sorbed P undergoes further transformations with time (Mattingly, 1975; Parfitt *et al.*, 1989). Such processes may involve recrystallisation (Barrow, 1983), solid state diffusion (Willett *et al.*, 1988), or multiple P pools which are not in immediate exchange with the solution (Fardeau and Jappe, 1980) or which have differing affinities to P (Kanabo *et al.*, 1978). A measurement of available inorganic P therefore needs to consider both the amounts and the rates of release of P from the solid phase. Among the approaches taken to evaluate release rates are repeated water extracts, and desorption isotherms (Bache and Williams, 1971; Fox and Kamprath, 1970), possibly at elevated temperatures to substitute for impractically long reaction times (Barrow and Shaw, 1975).

Isotopic dilution methods can be used to measure the size of the P pool from which solution P can be replenished over a few hours, which has been used to define an exchangeable or available P pool. A continuing disappearance of tracer from the soil solution after the initial rapid isotopic dilution indicates the existence of a "sink" for ^{32}P that consists of sorbing phases which are not in direct exchange with the soil solution (Fardeau and Jappe, 1980; Salcedo *et al.*, 1991). This sink represents an activity of less soluble or slower pools of soil P which may replenish available P at rates varying from days to years (Morel *et al.*, 1994). Parfitt *et al.* (1989) confirmed that the slow sorption reaction can reduce plant available P, but the process is also reversible and can supply P to plants (Tiessen *et al.*, 1993).

I.4 ORGANIC PHOSPHORUS TRANSFORMATIONS

In addition to inorganic processes, the turnover of organic matter constantly releases P into the solution from Po mineralisation. Through the mineralisation of Po, the pool size of "total available" P is strongly time dependent. In natural ecosystems where Po recycling is an integral part of plant P supply, plant available P is a functional concept, measurable only by quantifying litter production and mineralisation as well as soil P pools (Tiessen *et al.*, 1994a). Any fertiliser responses (for purposes such as pasture improvement) would be erratic. Therefore available P must be defined with respect to the type of ecosystem in which it is measured, taking into consideration the type of P sink, i.e. a plant, plant community or crop, the period over which P supply is required, such as a cropping season, year, or growth cycle, and the rates of recycling that occur during that period. The element of time in the definition of plant available P has not received much attention in the past, since efforts were concentrated on annual crops under arable agriculture. Agroforestry, alley cropping, shifting cultivation, managed rotations and fallows, pastures, and the concept of "sustainability" of agriculture, all require that understanding and analytical capabilities develop beyond the "immediately available" nutrient pool.

I.5 SEQUENTIAL EXTRACTIONS

A potentially available P pool, analogous to the mineralisable N or S pools measured with incubation and leaching techniques (Ellert and Bettany, 1988) is not feasible, because of the reactivity of solution P with the soil's mineral phase. This lead to the design of sequential fractionations which first remove labile P, and then the more stable forms. The sequential extraction of Chang and Jackson (1957) attempted to separate chemically identifiable forms of Pi, but problems of re-precipitation of extracted P, and the similarity in the solubilities of Fe and Al associated P have limited its success (Williams *et al.*, 1967; Williams and Walker, 1969). Significant amounts of organic P are extracted with the Chang and Jackson procedure, but have received little attention, although they have been shown to be important in plant nutrition (Kelley *et al.*, 1983).

A sequence of alkaline followed by acid extracts, which forms part of the Chang and Jackson scheme gives a reliable distinction between Al+Fe and Ca-associated Pi (Kurmies, 1972). This distinction is useful because the balance between primary Ca-associated and secondary Al+Fe Pi forms reflects the weathering stage of the soil and indicates the presence of Ca rock phosphate fertiliser in weathered soils that otherwise contain little Ca-bound P. In most unweathered soils P occurs predominantly in the form of apatites, calcium phosphates containing carbonate, fluoride, sulphate, hydroxide, and a number of different cations, or in primary minerals where P substitutes isomorphously for silicate in crystal structures. Weathering and leaching reduce total P contents, and cause the formation of secondary Fe and Al phosphates with low solubilities. At the same time, the accumulation of organic matter in the soil as a result of the establishment of vegetation is accompanied by the formation of organic P (Po) pools.

Using these concepts, Hedley *et al.* (1982a) attempted to quantify labile, immediately plant available Pi, Ca-Pi, Fe+Al-Pi, and labile and more stable forms of Po using the following extracts:

Resin and bicarbonate extractable Pi are the labile forms of Pi which are thought to consist of Pi adsorbed on surfaces of more crystalline P compounds, sesquioxides or carbonates (Mattingly, 1975). Hydroxide extractable Pi has lower (or slower) plant availability (Marks, 1977) and is thought to be associated with amorphous and some crystalline Al and Fe phosphates. A more precise characterisation of these Pi forms is usually not possible or desirable since mixed compounds containing Ca, Al, Fe, P and other ions predominate in soils (Sawhney, 1973). Labile bicarbonate extractable Po is easily mineralisable and contributes to plant available P over one growing season (Bowman and Cole, 1978). More stable forms of Po involved in the long-term transformations of P in soils are extracted with hydroxide (Batsula and Krivonosova, 1973). By comparing the P compositions of different field soils, each of these extracts was empirically assigned a role in natural P transformations associated with microbial uptake (Hedley *et al.*, 1982a), plant roots (Hedley *et al.*, 1982b), cultivation (Tiessen *et al.*, 1983), or soil development (Tiessen *et al.*, 1984; Roberts *et al.*, 1985). The residual, un-extractable P in a number of West African soils, amounting to 20 and 60% of total P, was positively correlated to both Fe and Al which are responsible for the long-term stabilisation of P (Tiessen *et al.*, 1991).

Sharpley *et al.* (1985) related the fractions obtained by this method to a number of different P tests on 86 different soils with representatives from most soil orders (Tiessen *et al.*, 1984). Resin extractable labile P was related to Bray I and Olsen P tests in calcareous soils, but the correlation decreased with increasing weathering stage of the soils. Conversely, the Mehlich I P test was a good predictor of labile P in highly weathered soils but did not perform well on younger soils (Sharpley *et al.*, 1985).

Although the sequential fractionation extracts several Po pools, their nature is even less well defined than that of the inorganic extracts (Stewart and Tiessen, 1987). The residue of the original fractionation of Hedley *et al.* (1982a) often contained significant amounts of organic P that sometimes participated in relatively short-term transformations despite its recalcitrance with respect to chemical extractants. Particulate organic debris which may be rapidly mineralised by soil fauna and flora usually contains P that is not extractable. On relatively young, Ca-dominated soils this residual organic and inorganic P can be extracted by a second NaOH treatment following the acid extraction (Condron *et al.*, 1990), leaving negligible amounts of P in the final residue. On more weathered soils, hot HCl (Metha *et al.*, 1954) extracts most of the organic and inorganic residual P (Condron *et al.*, 1990). The complete method is described by Tiessen and Moir (1993).

Despite the success of fractionating most of the soils P, it has been difficult to assign chemically extracted Po fractions specific roles in soil P transformations, because Po turnover frequently depends on the mineralisation of organic matter during which P is released as a side product. Most field observations of P mineralisation are not explained by the biochemical release of P, but by the biological turnover of organic matter with concomitant

P release (*cf.* McGill and Cole, 1981). Correlations have been used successfully to show the dependence of Pi turnover and stabilisation on soil mineral components such as non-crystalline Al and Fe, but this approach fails when relating labile P fractions to organic P or organic matter (e.g. Ayodele and Agboola, 1983), since P availability itself limits the accumulation of organic matter in many soils (Tiessen and Stewart, 1983).

Microbes may play an important role in the short-term cycling and availability of P by mechanisms such as uptake-storage-release, mineralisation and solubilisation (Lee *et al.*, 1990; Singh *et al.*, 1989; Parfitt *et al.*, 1989; McLaughlin *et al.*, 1988; Asea *et al.*, 1988), and microbial P has been determined as part of the Hedley *et al.* (1982a) fractionation. In some ecosystems such as tropical rainforests on very infertile soils, closed plant to plant or litter to plant recycling represent a major portion of the active P cycle (Medina and Cuevas, 1989), while any chemical soil fractionation entirely ignores the complexity of such biotic P transformations. Few complete ecosystem analyses of P cycles incorporating complex biological compartments have been done (Chapin, 1978; Tiessen *et al.*, 1994a).

I.6 SPATIAL VARIABILITY

In addition to the problems associated with the description of soil P in any one sample, and its transformations with time, the forms and transformations of P are highly variable in landscapes, profiles or even within horizons. This variability cannot be ignored when soil fertility, plant performance or nutrient cycling are to be evaluated. While field variability of total soil P may be low (CV < 20 %), resin extractable P and other labile fractions are very variable (CV > 50%) since they depend to a much greater extent on micro relief, soil moisture, clay content, plant uptake etc. (Tiessen and Santos, 1989). Mineralogical differences between horizons will affect P sorption, with low sorption in organic matter rich top soils and high sorption in sesquioxide-rich accumulation horizons (Stewart *et al.*, 1990). Similarly, the presence of highly sorbing lateritic materials in a soil profile will greatly modify P cycling (Tiessen *et al.*, 1991). Cultivation and erosion will translate such differences into lateral variability at field scales.

I.7 CURRENT APPROACHES AND DIRECTIONS

Short-term fertility assessments of P status on cultivated soils are successfully performed in many places with mild acid (Bray) or alkaline (Olson, Dabin) extractants. Extractions with anion exchange resins are easier to interpret since they do not modify the chemical composition of the soil solution. Results of such extractions can therefore more easily be compared across different soil types and ecosystems. Ion exchange resins are available on polymer membranes which can be cut to suitable size and used repeatedly for the extraction of soil P. Results obtained with these membranes are close but not absolutely comparable to results obtained with the more traditional resin beads. The use of resin membranes, though, is much easier because soil can be washed off easily, membranes are easily picked up, transferred from flasks, re-extracted, and re-used so that now resin extraction has become simple enough for routine applications, and such membranes have been adapted to the *in situ* measurement of nutrient anions (*cf.* Schoenau *et al.*, 1993). Simple extraction data can be supplemented with an evaluation of P fixation by equilibrating soil samples with different levels of P (Fox and Kamprath, 1970) to arrive at a reasonably close prediction of P requirements under conditions of annual arable agriculture.

The study of P transformations in mixed or perennial agricultural systems or in native ecosystems poses much greater problems. The questions addressed in such studies include: Is P limiting accumulation or transformations of C and N? What constraints does P availability impose on current and future crop production in continuous cropping or rotational agriculture? What are the mechanisms of regeneration of fertility under bush fallow? In each case the entire complex of Pi and Po transformations including mineral interactions and the turnover of organic matter needs to be studied. P sorption, Fe and Al activities, and P fractionations following the procedures of Kurmies or Hedley and co-workers have proved to be successful in elucidating Pi chemistry and transformations. These methods can follow the fate of Pi or fertiliser P with liming, incubations, exhaustive cropping or similar controlled manipulations. Correlative studies with other soil properties, along environmental gradients, with differences in cultivation or soil development have helped the understanding of the P cycle and promise further successes in new ecosystems.

In ecosystems where biotic P cycling through close recycling of plant litter is important, such as in nutrient-limited tropical rain forests, chemical P transformations in the soil are of course of limited relevance. In such ecosystems detailed analyses of ecosystem components such as leaf P, litter P, plant P, decomposition rates, root growth and root distribution and the role of bacteria and fungi must be evaluated in order to understand the P cycle. Such studies are complicated by the unavailability of a long-lived isotopic tracer for P. Only the use of repeated elemental budgets in different ecosystem components along seasons and the consideration of organic matter transformations have provided insight into the P cycle in such ecosystems (Medina and Cuevas, 1989; Tiessen *et al.*, 1994b).

Studies of the availability, cycling and chemical composition of Po suffer from even greater methodological limitations. Po fractionations are included in Hedley's approach but the biological functions of the chemically extractable fractions are not easily defined. While these fractions have changed predictably under manipulations such as incubations, fertilisations, manuring, exhaustive cropping, etc. an understanding of the role of Po in the environmental P cycle must be based on an understanding of organic matter transformations. Separation of recent litter, floatable organic matter, clay associated or mineral stabilised organic matter, and analysis of the associated P are methods that have shown some success. Short-term turnover of Po is largely microbially mediated and microbial P measurements using one of the chloroform methods are useful in this context. Phosphatase activity may also affect biological P availability and turnover (Oberson *et al.*, 1993). In highly P sorbing soils, the P released by chloroforming a microbial population is rapidly removed from extractable pools and the chloroform methods become relatively inaccurate and require large fudge factors. Nevertheless, microbial P measurements can be used in studies comparing similar soils and ecosystems.

P cycles in soils and entire ecosystems depend on the chemistry of Pi and the transformations of organic P and hence of organic matter. Since P often limits organic matter transformations quite complex interrelationships exist and conceptual (Tiessen *et al.*, 1984) or mathematical (Cole *et al.*, 1977) models may help in the analyses of soil and ecosystem data and in the planning of research approaches.

References

Adepetu, J.A. and Corey, R.B. (1976) Organic phosphorous as a predictor of plant-available phosphorus in soils of southern Nigeria. *Soil Science* **122**, 159-164.

Ae, N., Arihara, J., Okada, K., Yoshihara, T. and Johansen, C. (1990) Phosphorus uptake by pigeonpea and its role in cropping systems of the Indian subcontinent. *Science* **248**, 477-480.

Agboola, A.A. and Oko, B. (1976) An attempt to evaluate plant available P in western Nigerian soils under shifting cultivation. *Agronomy Journal* **68**, 798-801.

Agboola, A.A. and Ayodele, O.J. (1983) An attempt to evaluate plant available P in western Nigeria savannah soils under traditional fallow systems. *IMPHOS 3rd International Congress on Phosphorus Compounds*, Brussels, pp 261-287.

Amer, F., Bouldin, D.R., Black, C.A. and Duke, F.R. (1955) Characterisation of soil phosphorus by anion exchange resin adsorption and ^{32}P equilibration. *Plant and Soil* **6**, 391-408.

Anyaduba, E.T. and Adepetu, J.A. (1983) Predicting the phosphorus fertilisation need of tropical soils: significance of the relationship between critical soil solution P, requirement of cowpea, P sorption potential, and free iron content of the soil. *Beiträge zur tropischen Landwirtschaft und Veterinärmedizin* **21**, 21-30.

Asea, P.E.A., Kucey, R.M.N. and Stewart, J.W.B. (1988) Inorganic phosphate solubilization by two Penicillium species in solution culture and soil. *Soil Biology and Biochemistry* **20**, 459-464.

Ayodele, O.J. and Agboola, A.A. (1983) Evaluation of phosphorus in savannah soils of western Nigeria under bush fallow systems. *Journal of Agricultural Science* **101**, 283-289, Cambridge.

Bache, B.W. and Williams, E.G. (1971) A phosphate sorption index for soils. *Journal of Soil Science* **22**, 289-301.

Barrow, N.J. and Shaw, T.C. (1975) The slow reactions between soil and anions. 2. Effect of time and temperature on the decrease in phosphate concentration in the soil solution. *Soil Science* **119**, 167-177.

Barrow, N.J. (1983) On the reversibility of phosphate sorption by soils. *Journal of Soil Science* **34**, 751-758.

Batsula, A.A. and Krivonosova, G.M. (1973) Phosphorus in the humic and fulvic acids of some Ukrainian soils. *Soviet Soil Science* **5**, 347-350.

Bowman, R.A. and Cole, C.V. (1978) Transformations of organic phosphorus substances in soils as evaluated by $NaHCO_3$ extraction. *Soil Science* **125**, 49-54.

Bray, R.H. and Kurtz, L.T. (1945) Determination of total, organic and available forms of phosphorus in soils. *Soil Science* **59**, 39-45.

Chang, S.C. and Jackson, M.L. (1957) Fractionation of soil phosphorus. *Soil Science* **84**, 133-144.

Chapin, F.S., Barsdate, R.J. and Barèl, D. (1978) Phosphorus cycling in Alaskan coastal tundra: a hypothesis for the regulation in nutrient cycling. *Oikos* **31**, 189-199.

Condron, L.M, Moir, J.O., Tiessen, and Stewart, J.W.B. (1990) Critical evaluation of methods for determining organic phosphorus in tropical soils. *Soil Science Society of America Journal* **54**, 1261-1266.

Curtin, D., Syers, J.K. and Smillie, G.W. (1987) The importance of exchangeable cations and resin-sink characteristics in the release of soil phosphorus. *Journal of Soil Science* **38**, 711-716.

Dabin, B. (1967) *Methode Olsen modifié.* Cahiers ORSTOM-Pedologie 5-3.

Egnér, H., Riehm, H. and Domingo, W.R. (1960) *Untersuchungen über die chemische Bodenanalyse als Grundlage für die Beurteilung des Nährstoffzustandes der Böden. 2. Chemische Extraktionsmethoden zur Phosphor- und Kaliumbestimmung.* Kungl. Landbrukshögskolans Annaler, Uppsala, Sweden, 26, 199-215.

Ellert, B.H. and Bettany, J.R. (1988) Comparisons of kinetic models for describing net sulfur and nitrogen mineralization. *Soil Science Society of America Journal* **52**, 1692-1702.

Fardeau, J.C. and Jappe, J. (1980) Choix de la fertilisation phosphorique des sols tropicaux: emploi du phosphore 32. *Agronomie Tropicale* **35**, 225-231.

Fox, R.L. and Kamprath, E.J. (1970) Phosphate sorption isotherms for evaluating the phosphate requirements of soils. *Soil Science Society of America Proceedings* **34**, 902-907.

Hedley, M.J., Stewart, J.W.B. and Chauhan, B.S. (1982a) Changes in inorganic and organic soil phosphorus fractions induced by cultivation practices and by laboratory incubations. *Soil Science Society of America Journal* **46**, 970-976.

Hedley, M.J., White, R.E. and Nye, P.H. (1982b) Plant-induced changes in the rhizosphere of rape seedlings. III. Changes in L-value, soil phosphate fractions and phosphatase activity. *New Phytologist* **91**, 45-56.

Hughes, J.C. (1982) High gradient magnetic separation of some soil clays from Nigeria, Brazil and Colombia. I. The interrelationships of iron and aluminium extracted by acid ammonium oxalate and carbon. *Journal of Soil Science* **33**, 509-519.

Hughes, J.C. and Le Mare, P.H. (1982) High gradient magnetic separation of some soil clays from Nigeria, Brazil and Colombia. II. Phosphate adsorption characteristics, the inter-relationships with iron aluminium and carbon, and comparison with whole soil data. *Journal of Soil Science* **33**, 521-533.

Juo, A.S.R. and Fox, R.L. (1977) Phosphate sorption characteristics of some bench-mark soils of west Africa. *Soil Science* **124**, 370-376.

Kanabo, I.A.K., Halm, A.T. and Obeng, H.B. (1978) Phosphorus adsorption by surface samples of five ironpan soils of Ghana. *Geoderma* **20**, 299-306.

Kelley, J., Lambert, M.J. and Turner, J. (1983) Available phosphorus forms in forest soils and their possible ecological significance. *Communications in Soil Science and Plant Analysis* **14**, 1217-1234.

Kurmies, B. (1972) Zur Fraktionierung der Bodenphosphate. *Die Phosphorsaure* **29**, 118-151.

Le Mare, P.H. (1981) Exchangeable phosphorus, estimates of it from amorphous iron oxides, and soil solution phosphorus, in relation to phosphorus taken up by maize. *Journal of Soil Science* **32**, 285-299.

Le Mare, P.H. (1982) Sorption of isotopically exchangeable and non-exchangeable phosphate by some soils of Colombia and Brazil, and comparisons with soils of southern Nigeria. *Journal of Soil Science* **33**, 691-707.

Lee, D., Han, X.G. and Jordan, C.F. (1990) Soil phosphorus fractions, aluminum, and water retention as affected by microbial activity in an Ultisol. *Plant and Soil* **121**, 125-136.

López-Hernández, D. and Flores-Aguilar, D. (1979) La Desorcion de fosfatos en suelos. Implicaciones fisioecologicas en el proceso. *Acta Cient. Venezolana* **30**, 23-35.

Marks, G. (1977) Beitrag zur praezisierten Charakterisierung von Pflanzen verfuegbaren Phosphat in Ackerboeden. *Arch. Acker Pflanzenbau Bodenkd.* **21**, 447-456.

Mattingly, G.E.G. (1975) Labile phosphate in soils. *Soil Science* **119**, 369-375.

McGill, W.B. and Cole, C.V. (1981) Comparative aspects of C, N, S and P cycling through soil organic matter during pedogenesis. *Geoderma* **26**, 267-286.

McLaughlin, M.J., Alston, A.M. and Martin, J.K. (1988) Phosphorus cycling in wheat-pasture rotations. II. The role of the microbial biomass in phosphorus cycling. *Australian Journal of Soil Research* **26**, 333-342.

Medina, E. and Cuevas, E. (1989) Patterns of nutrient accumulation and release in Amazonian forests of the upper Rio Negro Basin. In: Proctor, J. (ed.), *Mineral Nutrients in Tropicale Forest and Savanna Ecosystems*, Blackwell Scientific, Oxford, UK.

Metha, N.C., Legg, J.O., Goring, C.A.I. and Black, C.A. (1954) Determination of organic phosphorus in soils: I. Extraction method. *Soil Sciènce Society of America Proceedings* **18**, 443-449.

Morel, C., Tiessen, H., Moir, J.O. and Stewart, J.W.B. (1994) Phosphorus transformations and availability due to crop rotations and mineral fertilization assessed by isotopic dilution. *Soil Science of America Journal* **58,** 1439-1445.

Nelson, W.L., Mehlich, A. and Winters, E. (1953) The development, evaluation, and use of soil tests for phosphorus availability. In: Pierre, W.H. and Norman, A.G. (eds.), Soil Fertilizer Phosphorus. *Agronomy* **4**, 153-188. American Society of Agronomy Madison, WI, USA.

Oberson, A., Fardeau, J.C., Besson, J.M. and Sticher, H. (1993) Soil phosphorus dynamics in cropping systems managed according to conventional and biological agricultural methods. *Biology and Fertility of Soils* **16,** 111-117.

Olsen, S.R., Cole, C.V., Watanabe, F.S. and Dean, L.A. (1954) Estimation of available phosphorus in soils by extraction with sodium bicarbonate. U.S. Department of Agriculture Circular 939.

Onken, A.B., Matheson, R. and Williams, E.J. (1980) Evaluation of EDTA-extractable P as a soil test procedure. *Soil Science Society of America Journal* **44**, 783-786.

Parfitt, R.L., Hume, L.J. and Sparling, G.P. (1989) Loss of availability of phosphate in New Zealand soils. *Journal of Soil Science* **40**, 371-382.

Roberts, T.L., Stewart, J.W.B. and Bettany, J.R. (1985) The influence of topography on the distribution of organic and inorganic soil phosphorus across a narrow environmental gradient. *Canadian Journal of Soil Science* **65**, 651-665.

Roche, P. (1983) Les méthodes d'appréciation du statut phosphorique des sols. Leur application à l'estimation des besoins en engrais phosphatés. *IMPHOS 3rd International Congress on Phosphorus Compounds*, Brussels, pp.165-194.

Roche, P., Calba, H. and Fallavier, P. (1983) Estimation du niveau de fertilité phosphorique des sols de Cote d'Ivoire et évaluation des besoins en engrais phosphatés. *IMPHOS 3rd International Congress on Phosphorus Compounds*, Brussels, pp.289-297.

Salcedo, I.H., Bertino, F. and Sampaio, E.V.S.B. (1991) Reactivity of fertilizer P in tropical soils by isotopic exchange kinetics of ^{32}P. *Soil Science Society of America Journal* **55**, 140-145.

Sawhey, B.L. (1973) Electron microprobe analysis of phosphates in soils and sediments. *Soil Science Society of America Proceedings* **37**, 658-660.

Schoenau, J.J., Qian, P. and Huang, W.Z. (1993). Assessing sulphur availability in soil using anion exchange membranes. *Sulphur Agric* **17,** 13-17.

Sharpley, A.N., Jones, C.A., Gray, C., Cole, C.V., Tiessen, H. and Holzhey, C.S. (1985) A Detailed Phosphorus Characterisation of Seventy-eight Soils. U.S. Department of Agriculture, *Agricultural Research Service Publication* **31**, 32 pp.

Singh, J.S., Raghubanshi, A.S., Singh, R.S. and Srivastava, S.C. (1989) Microbial biomass acts as a source of plant nutrients in dry tropical forest and savanna. *Nature* **338**, 499-500.

Stewart, J.W.B. and Tiessen, H. (1987) Dynamics of soil organic phosphorus. *Biogeochemistry* **4**, 41-60.

Stewart, J.W.B., Tiessen, H. and Frossard, E. (1990) Phosphorus requirements of semi-arid lands. In: Unger, P.W., Sneed, T.V., Jordan, W.R. and Jensen, R.W. (eds.), *Challenges in Dryland Agriculture - A Global Perspective*, Proceedings of the International Conference on Dryland Farming, Amarillo/Bushland, Texas, August 15-19, 1988, Texas Agricultural Experiment Station, Bushland, pp.433-435.

Tiessen, H., Abekoe, M.K., Salcedo, I.H. and Owusu-Bennoah, E. (1993) Reversibility of phosphorus sorption by ferruginous nodules. *Plant and Soil* **153,** 113-124.

Tiessen, H. and Carneiro dos Santos, M. (1989) Variability of C, N and P content of a tropical semiarid soil as affected by soil genesis, erosion and land clearing. *Plant and Soil* **119**, 337-341.

Tiessen, H., Chacon, P. and Cuevas, E. (1994a) Phosphorus and nitrogen status in soils and vegetation along a topsequence of dystrophic rainforests on upper Rio Negro. *Oecologia* **99,** 145-150.

Tiessen, H., Cuevas, E. and Chacon, P. (1994b) The role of soil organic matter in sustaining soil fertility. *Nature* **371,** 783-785.

Tiessen, H. and Stewart, J.W.B. (1983) The biogeochemistry of soil phosphorus. In: Caldwell, D.E., Brierley, J.A. and Brierley, C.L. (eds.), *Planetary Ecology*, Van Norstrand Reinhold Company, New York, pp.463-472.

Tiessen, H., Stewart, J.W.B. and Moir, J.O. (1983) Changes in organic and inorganic phosphorus composition of two grassland soils and their particle size fractions during 60-70 years of cultivation. *Journal of Soil Science* **34**, 815-823.

Tiessen, H., Stewart, J.W.B. and Cole, C.V. (1984) Pathways of phosphorus transformations in soils of differing pedogenesis. *Soil Science Society of America Journal* **48**, 853-858.

Tiessen, H., Frossard, E., Mermut, A.R. and Nyamekye, A.L. (1991) Phosphorus sorption, and properties of ferruginous nodules from semiarid soils from Ghana and Brazil. *Geoderma* **48**, 373-389.

Tiessen, H. and Moir, J.O. (1993). Characterisation of available P by sequential extraction. Chapter 10. In: M.R. Carter (ed.). Soil Sampling and Methods of Analysis, CRC Press, pp. 75-86.

Van der Zaag, P., Fox, R.L., De La Pena, R., Laughlin, W.M., Ryskamp, A., Villagarcia, S. and Westermann, D.T. (1979) The utility of phosphate sorption curves for transferring soil management information. *Tropical Agriculture (Trinidad)* **56**, 153-160.

Willett, I.R., Chartres, C.J. and Nguyen, T.T. (1988) Migration of phosphate into aggregated particles of ferrihydrite. *Journal of Soil Science* **39**, 275-282.

Williams, J.D.H. and Walker, T.W. (1969) Fractionation of phosphate in a maturity sequence of New Zealand basaltic soil profiles. *Soil Science* **107**, 22-30.

Williams, J.D.H., Syers, J.K. and Walker, T.W. (1967) Fractionation of soil inorganic phosphate by a modification of Chang and Jackson's procedure. *Soil Science Society of America Proceedings* **31**, 736-739.

Appendix J ISOTOPE STUDIES IN TROPICAL SOIL BIOLOGY

Roel Merckx

J.1 INTRODUCTION

The general interest in using (radio)isotopes to study several processes of relevance to nutrient cycling is mainly due to the relative ease with which a certain element can be followed through its different transformations. Detection of radioactivity, or increased abundance in the case of stable isotopes, in a given compartment of the ecosystem studied is an unequivocal proof of its origin. The most straightforward example is the amendment of radioisotope-labelled fertilisers to a given agroecosystem to assess such things as fertiliser efficiency, losses to various environmental compartments and turnover in soil.

Within the Tropical Soil Biology and Fertility Programme focus is on nutrient cycling in systems characterised by a high degree of organic residue recycling. Effective management of these residues will benefit from a detailed knowledge on the fate of the different nutrients they contain. Traditionnally, carbon, nitrogen and phosphorus are considered as key elements. Therefore, in this review, isotope studies related to the cycling of those three elements are emphasised.

Ignoring the very short-lived radioisotopes of nitrogen (e.g. ^{13}N with $^{1/2}T$ of 10 min) which are entirely useless in this type of ecological research, we can divide the isotope studies into two groups according to their detection: radioisotope studies relying on γ and/or β-spectroscopy and stable isotope studies using mass spectrometry. With respect to the latter, ^{15}N is the most frequently used isotope for nitrogen. (Techniques available to quantify nitrogen fixation using ^{15}N methodology are reviewed in Appendix G, "Techniques for quantifing nitrogen fixation".)

Focus here will be on the methodology associated with the use of labelled residues to study their decomposition and cycling of their elements within a given system and on methods estimating root material production and turnover.

Note: A general methods overview can be found in Hardarson, G. (Ed.) (1990) Use of Nuclear Techniques in Studies of Soil-Plant Relationships. Training Course Series No 2, International Atomic Energy Agency (IAEA), Vienna, Austria.

J.2 DOUBLE (^{14}C, ^{15}N) OR TRIPLE LABELLED (^{14}C, ^{15}N, ^{32}P) RESIDUES

Decomposition experiments with uniformly labelled residues have been performed at various places and with various plant species (Jenkinson, 1981; Paul and van Veen, 1978; Ladd and Martin, 1984; Scharpenseel and Neue, 1984). Nevertheless the number of species involved is rather limited and certainly fails to adequately cover the huge range of materials of relevance to tropical farming systems in general and to agroforestry-related systems in particular.

The general idea of decomposition experiments with labelled residues being the detection of residue-derived elements in various compartments of the system and how this evolves with time, the first question should be for how long these measurements will have to be continued. This is critical as it predetermines the level of radioactivity to be added to the system. A general principle in radioisotope methodology is the so called "ALARA" principle recommending that radioactivity amounts should be As Low As Reasonably Achievable (ICRP, 1977). As such, every calculation starts with the detection limits of the counting device and then calculating back with several assumptions on transfer processes of the isotope between the different compartments involved. An example of such calculation is given below. Furthermore, the length of the experimental period also determines what kind of isotopes should be used. The use of ^{32}P and ^{33}P is restricted to rather short-term experiments in view of their short half-lives (14.3 days and 25.3 days respectively). Nevertheless, ^{32}P and ^{33}P labelling of residues has been used to determine how much of the phosphorus present in medic residues was recovered by a wheat crop in South Australia (McLaughlin *et al.*, 1988).

J.3 LABELLING THE PLANT MATERIAL

This is undoubtedly the bottleneck of the experiments with labelled residues. Although labelling with ^{15}N, ^{32}P can be achieved relatively easy using nutrient solution techniques (Ladd *et al.*, 1981; McLaughlin *et al.*, 1988), labelling with ^{14}C requires a relatively sophisticated growth chamber, preferably with control of the specific activity of the atmosphere.

For all three labels it is mandatory to ensure homogeneous labelling, i.e. that the specific activity of the element is constant amongst the different fractions of the material. Provision of ^{14}C at constant specific activity requires a closed canopy under which plants can grow and where the radioactivity of the atmosphere can be monitored and kept at a preset level. The most economical way of achieving this is by supplying radioactive $Na_2{}^{14}CO_3$ into an acid bath at rates dictated by a preset radioactivity level in the growth chamber as monitored by a G.M. tube (Gurr and Adey, 1978). Homogeneous labelling with ^{15}N or $^{32}P/^{33}P$ can be acheived by (a) growing the plants in a medium contributing minimally to the N or P demand of the plant and (b) amending the nutrients distributed according to the plant needs over the growing season. For nitrogen, ^{15}N labelled $(NH_4)_2SO_4$ can be used, whereas for phosphorus ^{32}P labelled $Ca(H_2PO_4)_2$ or ^{33}P labelled $Ca(H_2PO_4)_2$ can be used. By a proper selection of the nutrient additions according to earlier established growth rates, homogeneous labelling is encouraged. With respect to the use of phosphorus isotopes, the short half-lifes of both (^{32}P, 14.3 days; ^{33}P, 25.3 days) restrict their use to short growing periods.

J.4 DECOMPOSITION EXPERIMENTS

No matter how the number of growing facilities of labelled plant materials will increase in the future, the relatively high cost of the labelled materials will limit their use to small-scale, more detailed experiments. In general, the material can be used in the field if confined to small surfaces, or in small litter bags. The material can be used as a surface litter or incorporated in the topsoil depending on the experimental objectives but again the limited amount available will require that some fragmentation of the residues has to be done to achieve sufficient homogeneity within one cylinder. Application rates and modes of application can vary according to the research needs.

Procedure

A typical experiment may offer two main options: (i) closed cores restricting root in-growth; or (ii) mesh type cores (TSBF, 1991) allowing free horizontal passage of roots and fauna. Effectively, the core only serves the purpose of confining the litter to a specified surface. A wide-mesh cover prevents the wind from removing residues from the surface. The procedure described by Ladd *et al.* (1983) is generally applicable:

Edged, open-ended plastic cylinders (17.5 cm long by 10.3 cm diameter) are driven to 15 cm depth in soils at predetermined field sites. At each site cores, in numbers determined by the experimental objectives, are randomized within replicated (2 or 3) blocks. Soils, to say 7.5 cm depth, are removed from each cylinder, weighed, sieved (2 mm), mixed and bulked for each block at the sites.

Soil subsamples, of a weight equal to that of the average weight removed from each cylinder, are mixed with preweighed ^{14}C-labelled plant materials, of a radioactivity set by the duration of the experiment and type of analyses (see Example 1 below). The soils are returned to the cylinders, and the surfaces tamped gently so that the bulk density of the amended soil is the same as that of soil prior to its removal.

Later, at appropriate times and especially in the first year of decomposition, replicated cylinders are removed from the field, and the soils from each are dried and sieved (2 mm) *in toto*. Subsamples (20 g) are then more finely ground in a Tema mill before combusting aliquots (usually 0.5 g) by the procedure of Amato (1983): the soil is mixed with combustion liquid (6 ml) in Tecator tubes (75 ml), stoppered, then heated in a block at 130°C for 1 hour. The evolved $^{14}CO_2$ is trapped in ethanolamine (1 ml) contained in a glass vial and supported in the combustion tube (but away from the heated areas) by a bent glass rod. Routinely, absorption of $^{14}CO_2$ is allowed to take place overnight following cooling.

The vials containing ethanolamine with absorbed $^{14}CO_2$ are transferred to scintillation vials, shaken with scintillation fluid and the radioactivity determined. Corrections are made for background, and combustion and counting efficiencies, by using combusted standards.

Example 1. Calculation to determine the amount of radioactivity which needs to be added at the beginning of a 2-year experiment

To study the dynamics of the soil microbial biomass carbon in soil cores (cylinders), one wants to follow the ^{14}C-content of the microbial biomass during the 2 years following an amendment of ^{14}C-labelled residues; one must therefore add sufficient ^{14}C to each cylinder at the begining of the experiment to have sufficient ^{14}C in the microbial biomass for analysis after 2 years.

Assumptions for this example:

1. Organic ^{14}C residue decomposes at a rate similar to that observed in Nigeria for ryegrass (Jenkinson and Ayanaba, 1977). This means that after 2 years 14% of the initial input remains.

2. After comparable decomposition intensities the proportion of this residue present in soil microbial biomass is ± 5% (Ladd *et al.*, 1985).

3. Microbial biomass ^{14}C is measured by a fumigation extraction method (Ladd and Amato, 1988) where 50 g soil samples are extracted in 125 ml M KCl before and after 10 days fumigation with $CHCl_3$ at 25°C. The ^{14}C activity in these extracts are determined with liquid scintillation counting in a 1 ml sample. Microbial biomass ^{14}C is estimated from the gain in extractable ^{14}C after fumigation, multiplied by 3.25.

4. To obtain a good compromise between counting time and accuracy at least 100 counts per minute are needed, or 1.67 cps. With a counting efficiency of 83% this means that 2.0 Bq of ^{14}C in 1 ml extract is required.

Calculations

Amount of ^{14}C needed in the soil microbial biomass 2 years after amendment per 50 g soil:125 ml extractant = 2 Bq/ml x 125 ml x 3.25 = 0.81 kBq

Amount of soil in the top 7.5 cm of each cylinder = 625 cm^3 = 700 g
(assuming a bulk density of ± 1.1 g/cm^3)

Therefore the ^{14}C needed in the 700 g soil sample after 2 years = 0.81 kBq x (700/50) = 11.40 kBq ^{14}C

Given that only 5% of the initial ^{14}C is in the microbial biomass (Assumption 2), the requirement per cylinder (700 g soil) = 11.40 kBq x (100/5) = 0.228 MBq ^{14}C

But, given this is only 14% (Assumption 1) of the amount added as a residue, the amount needed to be added to each cylinder (700 g soil) 2 years earlier

= 0.228 MBq x (100/14) = 1.625 MBq ^{14}C

If a residue is added at a rate of 500 mg/cylinder (= 200 mg C/cylinder; assume organic materials contain 40% C) we need to apply materials with a specific activity of

1.625/200 = 8.13 kBq/mg C

The example serves as an illustration of how to determine the amounts of radioactivity needed to meet the experimental objectives; nevertheless, always keep radioisotope levels at a minimum (ALARA).

J.5 ^{13}C METHODOLOGY IN ESTIMATING TURNOVER RATES OF SOIL ORGANIC MATTER FRACTIONS

The method basically exploits the phenomenon that C_3 plants differ from C_4 plants in their natural abundance of the ^{13}C isotope due to their different photosynthetic pathways. Soil organic matter derived from plant residues therefore reflects its origin since this ^{13}C ratio is preserved during decomposition. Soil organic matter derived from a C_3 vegetation has typical mean $\delta^{13}C$ values of -27 ‰ whereas C_4 residues have $\delta^{13}C$ values of -12 ‰ (Smith and Epstein, 1971).

In situations where a sudden shift from C_3 to C_4 vegetation took place as for instance after forest clearing to sugar cane plantations, the gradual change in $\delta^{13}C$ of the different soil organic matter fractions allows one to calculate their respective decomposition or turnover rates (Balesdent *et al.*, 1990).

The method however has two main limitations:

1 its use in its ideal form is restricted to pure C_3 - C_4 transitions to maximize the difference in $\delta^{13}C$ values; and

2 access to a very accurate mass spectrometer is essential.

Isotope ratios are usually expressed as $\delta^{13}C$ values:

$$\delta^{13}C\ (‰) = \{(R\ sample\ /\ R\ reference) - 1\} \times 1000$$

where R is the ^{13}C:^{12}C ratio for sample and reference respectively.

To calculate the proportion (f) of organic carbon originating from source A in a mixture A + B whose $\delta^{13}C$ is equal to an experimentally determined δ, the following approximate formula is used:

$$\delta = f\delta_A + (1 - f)\ \delta_B$$

where δ_A and δ_B are the $\delta^{13}C$ values of the two sources A and B, in our case C_3 and C_4 derived residues respectively.

Usually, we know the total carbon content (C) of the soil (or fraction) so that:

$$C = C_A + C_B$$

where

$$C_A = f \times C$$

$$C_B = (1 - f)\ C$$

Therefore we can write:

$$\delta = (C_A \times \delta_A)\ /\ C\ +\ \delta_B\ -\ (C_A \times \delta_B)\ /\ C$$

or

$$\delta = (I / C) \times (\delta_A - \delta_B) C_A + \delta_B$$

with C_A, the amount of carbon derived from the source A (i.e. the C_3 vegetation) as the only unknown.

J.6 ^{14}C METHODS TO QUANTIFY ROOT PRODUCTION AND TURNOVER

Despite the substantial improvements in automated root observation techniques (see ASA, special publication no. 50) considerable difficulty persists in the translation of the data obtained into nutrient pools and turnover rates. Labelling of roots and the associated products of rhizodeposition has been achieved by exposing plant tops either continuously or at short intervals to an atmosphere containing $^{14}CO_2$ (Martin, 1977; Keith *et al.*, 1986). Methodology is adequately described in the relevant papers and it appears that for field measurements pulse-labelling techniques provide the most convenient compromise. Whereas continuous labelling requires large and sophisticated growth chambers, which are not easily accommodated in the field, pulse labelling can be performed with relatively simple equipment (Keith *et al.*, 1986; Swinnen *et al.*, 1994). However, the bulk of literature refers to annual crops like wheat, maize, soy bean, with relatively short growing periods. Larger plants (like trees in agroforestry systems) with longer growing periods have not been subject to these measurements for obvious reasons.

References

Amato, M. (1983) The determination of carbon ^{12}C and ^{14}C in plant and soil. *Soil Biology and Biochemistry* **15**, 611-612.

ASA Special Publication no. 50. (1987) *Minirhizotron observation tubes: methods and applications for measuring rhizosphere dynamics.* Taylor, H.M. (ed.), American Society of Agronomy, Crop Science Society of America, Soil Science of America.

Balesdent, J., Mariotti, A. and Boisgontier, D. (1990) Effect of tillage on soil organic carbon mineralization estimated from ^{13}C abundance in maize fields. *Journal of Soil Science* **41**, 587-596.

Gurr, C.G. and Adey, R.F.J. (1978) A device to control radio-carbon levels in the atmosphere of a plant growth cabinet. *Laboratory Practice* **27**, 197-199.

ICRP (1977) Recommendations of the ICRP, ICRP no. 26, Pergamon Press, Oxford.

Jenkinson, D.S. (1981) The fate of the plant and animal residues in soil. In: Greenland, D.J. and Hayes, M.H.B. (eds.), *The Chemistry of Soil Processes.* John Wiley and Sons, New York, pp.505-561.

Jenkinson, D.S. and Ayanaba, A. (1977) Decomposition of carbon-14 labelled plant material under tropical conditions. *Soil Science Society of America Journal* **41**, 912-915.

Keith, H., Oades, J.M. and Martin, J.K. (1986) Input of carbon to soil from wheat plants. Soil Biology and Biochemistry **18**, 445-449.

Ladd, J.N. and Martin, J.K. (1984) Soil organic matter studies. In: L'Annunziata, M.F. and Legg, J.O. (eds.), *Isotopes and Radiation in Agricultural Sciences* **I**. Academic Press, New York, pp.67-98.

Ladd, J.N. and Amato, M. (1988) Relationships between biomass ^{14}C and soluble organic ^{14}C of a range of fumigated soils. *Soil Biology and Biochemistry* **20**, 115-116.

Ladd, J.M., Oades, J.M. and Amato, M. (1981) Microbial biomass formed from ^{14}C, ^{15}N-labelled plant material decomposing in soils in the field. *Soil Biology and Biochemistry* **13**, 119-126.

Ladd, J.N., Jackson, R.B., Amato, M. and Butler, J.H.A. (1983) Decomposition of plant material in Australian Soils. I. The effect of quantity added on decomposition and on residual microbial biomass. *Australian Journal of Soil Research* **21**, 563-570.

Ladd, J.N., Amato, M. and Oades, J.M. (1985) Decomposition of plant material in Australian Soils. III. Residual organic and microbial biomass C and N from isotope-labelled legume material and soil organic matter, decomposing under field conditions. *Australian Journal of Soil Research* **23**, 603-611.

Martin, J.K. (1977) Factors influencing the loss of organic carbon from wheat roots. *Soil Biology and Biochemistry* **9**, 1-7.

McLaughlin, M.J., Alston, A.M. and Martin, J.K. (1988) Phosphorus cycling in wheat-pasture rotations. I. The source of phosphorus taken up by wheat. *Australian Journal of Soil Research* **26**, 323-331.

Paul, E.A. and van Veen, J.A. (1978) The use of tracers to determine the dynamic nature of organic matter. *Proceedings of the 11th International Congress of Soil Science* **3**, 1-43.

Scharpenseel, H.W. and Neue, H.V. (1984) The use of isotopes in studying the dynamics of organic matter in soil. In: *Organic Matter and Rice.* IRRI, Los Banos, Philippines, pp.273-309.

Smith, B.N. and Epstein, S. (1971) Two categories of ^{13}C / ^{12}C ratios for higher plants. *Plant Physiology* **47**, 380-384.

Swinnen, J., van Veen, J.A. and Merckx, R. (1994) ^{14}C pulse-labelling of field-grown spring wheat: an evaluation of its use in rhizosphere carbon budget estimations. *Soil Biology and Biochemistry* **26,** 161-170.

TSBF (1991) Core experiments, draft protocols version 1.0. TSBF, Nairobi.

Appendix K USING GROWTH ANALYSIS TO ESTIMATE PLANT NUTRIENT UPTAKE

Robert Scholes

K.1 INTRODUCTION

Growth analysis is a procedure for the analysis of data obtained by the sequential sampling of plant populations, individual plants or plant components. It reduces the data to a set of standardised quantities for comparison within and between experiments. Although originally developed to quantify carbon assimilation, it is equally applicable to nutrient uptake problems. Within the TSBF programme it represents one approach to estimating the time and magnitude of plant nutrient demand. The minimum data required to apply growth analysis is the dry mass of plant material harvested at two times, which are a known period apart. For nutrient applications the nutrient concentration in the tissues at each time must also be known. Typically, samples are taken about five times during the crop growth period, yielding four sets of growth analysis statistics.

The free availability of computers in recent years has had a considerable impact on the way in which the technique is applied. In its "classical" form, which has remained essentially unchanged for fifty years, the growth of the plant is assumed to be exponential; in other words, by a constant proportion of the already existing tissue at each time step. This assumption allows the use of relatively simple equations for estimating the instantaneous growth rate, but may be erroneous during certain phases of plant growth, or when growth is under strong external limitation. Precision of estimation relies on taking a large number of replicate samples at each of a few time intervals during the growth of the plant.

The more modern "functional" approach makes no restrictive *a priori* assumptions about the form of the growth curve. Fewer plant samples are taken, but at more times, and a smooth empirical function is fitted to the data. The growth rate is computed from the differential of this function. The functional approach is recommended to anyone with access to a desktop computer.

A simple introduction to growth analysis is given by Hunt (1978), and a more detailed treatment of the functional approach by Hunt (1982). The standard reference to the classical approach is Evans (1972). The approach as it relates to nutrient uptake problems is developed by Nye and Tinker (1977) and Barber (1984).

K.2 THE CLASSICAL APPROACH

The absolute rate of growth of a plant (g/day) changes continuously during the development of the plant. Growth analysis seeks to do three things:

(i) estimate the mean instantaneous growth rate over a period by sequential measurements of the absolute mass. The instantaneous growth rate is the slope of the plant mass versus time curve at any one moment (dW/dt);

(ii) express the growth rate relative to the amount of tissue present at the time of the measurement, so that comparisons can be made between different measurement periods and different plants. The basis of most growth analysis functions is an equation of the form (1/W) (dW/dt); and

(iii) apportion the observed growth into the contributions due to changes in size of the growing organs and changes in efficiency of existing organs, to assist in the interpretation of the growth process.

If ${}_1W$ is the dry mass of tissue harvested at ${}_1t$, and ${}_2W$ is the dry mass at time ${}_2t$, then if the growth of the plant is exponential, the mean Relative Growth Rate (RGR) over the period is approximated by

$${}_{1\text{-}2}RGR\ (g/g/day) = (\ln_2 W - \ln_1 W) / ({}_2t - {}_1t)$$

(For an explanation of notation used see Note 1 below.)

The efficiency of the photosynthetic process can be estimated by expressing the growth rate relative to the area of leaf available for photosynthesis, (L_A). This quantity is know as the Unit Leaf Rate, ULR, and its mean value over a period is

$${}_{1\text{-}2}ULR\ (g/m^2/day) = \{({}_2W - {}_1W) / ({}_2t - {}_1t)\} \times \{(\ln_2 L_A - \ln_1 L_A) / ({}_2L_A - {}_1L_A)\}$$

The mean size of the photosynthetic organs over a period, known as the Leaf Area Ratio, LAR, is given by

$${}_{1\text{-}2}LAR\ (m^2/day) = \{({}_1L_A/{}_1W) + ({}_2L_A/{}_1W)\} / 2$$

It can be shown that

$${}_{1\text{-}2}RGR = {}_{1\text{-}2}ULR \times {}_{1\text{-}2}LAR.$$

When these concepts are developed for a crop rather than for a single plant, and are related to the area of ground on which the crop is growing, an analogous set of quantities is defined, including the Crop Growth Rate, CGR, and the well known Leaf Area Index. The above calculations are usually (but not necessarily) performed on above-ground plant parts; other relationships, such as the Root:Shoot ratio relate above-ground growth to total growth and growth partitioning.

K.3 APPLICATION TO NUTRIENT UPTAKE

If the nutrient content of the plant is used in place of the plant dry weight, and the plant root system is treated as the assimilatory organ instead of the leaf area, then a set of equations describing below ground growth and nutrient uptake can be developed. These are analogous to those presented for above-ground carbon assimilation above.

If LR is the length of the root system (m), (see Note 2 below) and U is the nutrient content of the plant (U = C x W, where C is the nutrient concentration in the plant tissues), then the mean Root Extension Rate, RER, is

$${}_{1\text{-}2}RER \text{ (cm/cm/day)} = (\ln_2 L_R - \ln_1 L_R) / ({}_2t - {}_1t)$$

and the mean Relative Uptake Rate, RUR, is

$${}_{1\text{-}2}RUR \text{ (mol/mol/day)} = (\ln_2 U - \ln_1 U) / ({}_2t - {}_1t)$$

The mean net inflow of nutrients per unit of root length, I, which could be considered to be a root efficiency term analogous to the ULR, or alternatively as a measure of the availability of nutrients to the plant, is given by

$${}_{1\text{-}2}I \text{ (mol/m}^2\text{/day)} = \{({}_2U - {}_1U) / ({}_2L_R - {}_1L_R)\} \times \{(\ln_2 L_R - \ln_1 LR) / ({}_2t - {}_1t)\}$$

Because the partitioning of nutrients between roots and shoots is not equal, and considerable retranslocation of nutrients occurs within the plant, it is necessary to estimate the mass and nutrient concentration of all the plant parts, not just the above-ground portions.

Notes

1 The notation traditionally used with growth analysis is rather peculiar, but is followed here to avoid confusion should reference be made to other sources. The preceding subscript refers to the time of measurement, and the following subscript to the component which was measured. For example, ${}_3W_{leaf}$ would be the leaf dry mass at the time of the third sample.

2 Since the fine roots which are responsible for the nutrient uptake are cylinders with a diameter much smaller than their length, the length of the root system is a better measure of its assimilatory capacity than either the root surface area or root mass. It is also easier to measure that root surface area.

References

Barber, S.A. (1984) *Soil Nutrient Bioavailability: A Mechanistic Approach.* Wiley, New York.

Evans, G.C. (1972) *The Quantitative Analysis of Plant Growth.* Blackwell Scientific Publications, Oxford.

Hunt, R. (1978) *Plant Growth Analysis.* Studies in Biology 96. Edward Arnold, London.

Hunt, R. (1982) *Plant Growth Curves: the Functional Approach to Plant Growth Analysis.* Edward Arnold, London.

Nye, P.H. and Tinker, P.B. (1977) *Solute Movement in the Soil Root System.* Blackwell Scientific Publications, Oxford.

Appendix L USE OF THE CENTURY PLANT-SOIL ENVIRONMENTAL SIMULATION MODEL

Paul Woomer

The Tropical Soil Biology and Fertility Programme is currently evaluating the use of environmental simulation models as a means of testing our current perceptions of mechanisms operative within soils and their possible interactions. At the same time, we are actively calibrating and validating these models in order that longer-term consequences of land management strategies may be predicted. The model most frequently used to test soil organic matter dynamics and the effects of land use is the CENTURY model (Parton *et al.*, 1987, 1989). CENTURY is a general model of plant-soil ecosystems that may be used to represent many different land uses including natural and managed grasslands, natural and plantation forests and field monocrops. CENTURY simulates the dynamics of carbon, nitrogen, phosphorus and sulphur within plants, residues and soils. This model is available in PC format from Dr W J Parton of the Natural Resource Ecology Laboratory, Colorado State University, Fort Collins, CO 80523, USA for US$150.00.

L.1 COMPUTER SIMULATION OF DECOMPOSITION AND SOIL ORGANIC MATTER TRANSFORMATIONS

Detailed knowledge of decomposition rates of various fractions of applied material and soil organic matter, and the transformations between those fractions over time are best utilized through the development of computer models. Computers allow for iterative calculation of organic matter dynamics at a fixed time step. We are no longer dealing with "batch" decomposition, but detailed transformation between and loss from organic matter functional pools. Early models of this type were developed by Jenkinson and Rayner (1977) and Van Veen and Paul (1981).

Environmental simulation offers great opportunities in the anticipation of the effects of various land management strategies. While model output will never replace on-site experimentation, a large number of candidate land-use strategies can be rapidly assessed allowing for a pre-selection of experimental treatments and anticipation of experimental outcomes. Prior to this reliance upon modelling for these purposes, however, model output must be calibrated with, and validated by measured parameters. The Tropical Soil Biology and Fertility Programme (TSBF) is currently investigating the usefulness of the CENTURY model (Parton *et al.*, 1987, 1989) across a range of tropical ecosystems and land uses. Ideally, a validated model could be used in predictive mapping of the tropics using georeferenced databases and geographic information systems as has been done in temperate regions (Burke *et al.*, 1990).

The CENTURY model simulates plant productivity and soil dynamics for many land uses and management regimes of terrestrial ecosystems (Parton *et al.*, 1987). Early versions of

CENTURY accurately simulated productivity and soil development and nutrient accumulation in some tropical ecosystems (Parton *et al.*, 1989).

L.2 MODEL STRUCTURE AND DESCRIPTION

Parton *et al.* (1987) reported a detailed description of CENTURY; interested readers are referred to that publication. CENTURY was first developed to simulate grassland ecosystems of the Great Plains region of the US, and is currently being evaluated in many other areas of the world. Briefly, CENTURY simulates soil dynamic based on the dynamics of functional pools. These pools are differentiated on residence time in soils and the C:mineral ratios of these pools. Plant productivity is simulated via user-defined (or default) algorithms appropriate to a particular ecosystem and climate.

The overall structure of CENTURY is presented in Figure L.1. CENTURY consists of 3 major components: INITPAR, CENTURY and VIEW. INITPAR is used to develop site files that initialize the simulation. INITPAR may also be used to view model input and output definitions. The CENTURY component runs the environmental simulation; CENTURY must always be run with a data and fix file (i.e. site.DAT and site.FIX, respectively). The data file (.DAT) defines specific soil and climate attributes, the .FIX file is based on more general ecosystem attributes. CENTURY may be run for a grassland, cropland or forest, with various management options available to the user. The model output is then directed to the VIEW component. VIEW is the output module of TIME-ZERO: an integrated modelling environment. VIEW allows for the output to be plotted, printed, or saved. VIEW also allows the user to change input parameters and to re-run the simulation without re-entering INITPAR.

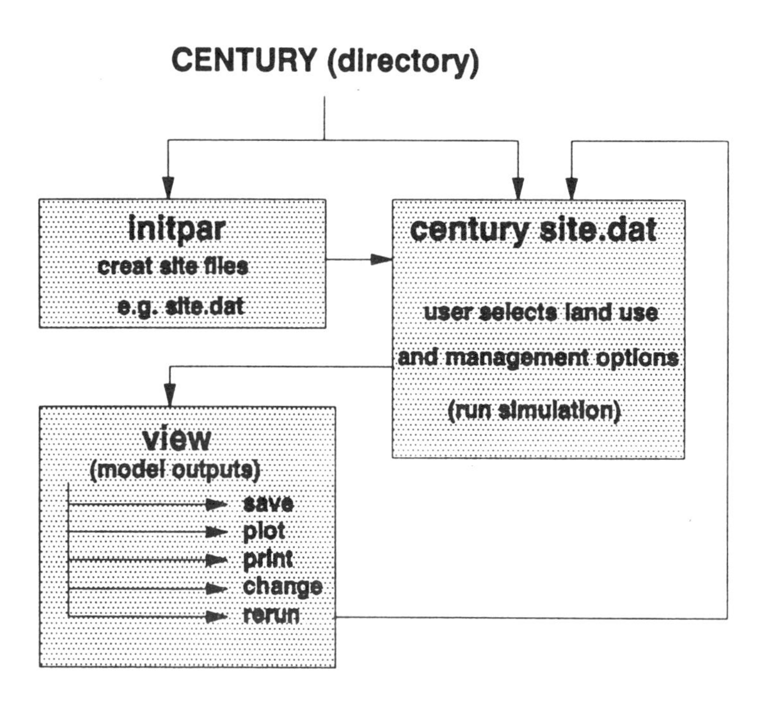

Figure L.1 General structure of CENTURY. Files are constructed in INITPAR, run in CENTURY and visualised in VIEW.

L.3 HARDWARE REQUIREMENTS AND INSTALLATION

The PC version of CENTURY is written in Fortran and requires an IBM PC or compatible with at least 512K RAM. While CENTURY can run on older processors such as an 80086, the simulations are quite time consuming. It is recommended that users run CENTURY on either a 80286 or 80386. For high resolution screen output, a VGA monitor is recommended although VIEW is supported by CGA and EGA monitors as well.

Together CENTURY and VIEW occupy 614K of hard disk drive. VIEW must be loaded as a subdirectory of the directory where CENTURY model files are stored. CENTURY site files (.DAT) may be stored directly in the main CENTURY directory, furthermore, site files cannot be read from an external drive.

L.4 RUNNING CENTURY

A session with the CENTURY model is always initiated by entering either INITPAR, if the user's intent is to create a new site.DAT file or CENTURY site.DAT if the user plans to run a simulation with existent site files. INITPAR is a user friendly routine whereby site files are constructed or modified. To enter INITPAR, the user must always refer to an already existent site.DAT file, modifying this file until representative of a new environment or land use. INITPAR is separated into minimum and complete initialising variables. After entering new variables through initpar, the user must save the file under a new name as CENTURY does not overwrite existing files. The initialising variables, user management options and suggested calibration and validation data required for the adoption of the CENTURY model into the Tropical Soil Biology and Fertility Programme are presented in Table L.1.

General information on ecological attributes is contained within the site.FIX files. For example, this file sets limits on the relationship between precipitation and plant productivity or on the rates of plant nutrient complexation with soil minerals. A HUMID.FIX file has been developed at TSBF headquarters that is universally used for sites of the lowland humid tropics with predominantly oxide soil mineralogies. These site.FIX files are more difficult to create than site.DAT files, requiring the use of a line editor.

A site data file must always accompany the CENTURY command (e.g. CENTURY SITE.DAT). Users are than asked to select a land use, pattern of events, a site.FIX file and management inputs. Several options of climate inputs are available, although the climate records contained within the site.DAT files are the most conveniently used. A CENTURY simulation is then generated for those conditions and the results of that simulation are automatically accessed through VIEW.

L.5 PRINCIPAL MODEL ROUTINES

CENTURY simulates the carbon, nitrogen, phosphorus and sulphur dynamics for grasslands, forests and cropping systems. These different plant production systems are linked to a common organic matter submodel that consists of 3 functional pools based on the residence

Table L.1 Minimum data requirements for a site file to initialise CENTURY, management options available to the user and examples of useful data for model calibration and validation.

Minimum file requirements:

clay content, number of elements simulated (C, N, P, S), time steps per month, soil pH, monthly precipitation, sand content, silt content, length of simulation (years), monthly minimum temperature, monthly maximum temperature, grass growth season (months of growth and senescence), crop growth season (months of growth, harvest and senescence), forest growth season (months of growth and litter fall).

Land use management options:

grass removal pattern (undisturbed, fire, grazed or both), grassland fire event (years, month), fertilisation timing and quantity (N, P, S, amount and when applied), addition of straw/residues (quantities, frequency, C:N, C:P, C:S and lignin content), crop/fallow pattern (years of various land use), tillage pattern (months, depth), timing of forest removal/return events, biomass allocation of forest removal events.

Useful calibration/validation data:

total soil organic (C, N, P), extractable nitrate and ammonium, soil microbial biomass (C, N, P), annual above-ground plant productivity, plant components of yield, above-ground standing biomass, standing dead biomass surface litter biomass, root biomass.

time of organic materials in soils. The plant nutrient subroutines (N, P and S) are incorporated into plant productivity and organic matter dynamics through user-defined C:nutrient ratios that float within certain limits. The dynamics of these plant nutrients is centred about a labile form that is available to plant uptake, subject to complexation with soil minerals or loss from leaching or erosion.

An example of the overall organic matter C and N dynamics for a crop-soil system is presented in Figure L.2. Plant productivity is controlled by moisture availability, temperature and the maximum and minimum C:nutrient ratios of new plant growth. User-defined partitioning coefficients determine the relative proportions of grain, shoots and roots. Upon senescence, plant materials enter the litter pools that are divided between easily metabolised and more recalcitrant materials. These litter forms enter into the different soil organic matter fractions at rates determined by their chemical compositions, soil textures and soil microclimate. System losses primarily occur as respiration of CO_2, and the nutrients associated with the respired C are assumed to become mineralised and enter their respective labile pools.

L.6 CENTURY OUTPUTS

The results of CENTURY simulation may be output as either line graphs or tables. CENTURY line graphs may contain up to six variables on the y axis. Users may select from a number of line styles and colours, and can create titles and legends within view, then output these directly to a printer or capture the image using an external grab feature. Graphs may also be stored as files within VIEW. Tables are created using the PRINT routines. The column headings may be renamed (e.g. different from their respective FORTRAN codes) or be products of combined output variables. These tables may be either printed or saved as data files importable into a number of commercially available spreadsheet and graphics programmes. The complete results of a simulation may also be stored within CENTURY for future use but these files occupy nearly as much disk space as the total programme.

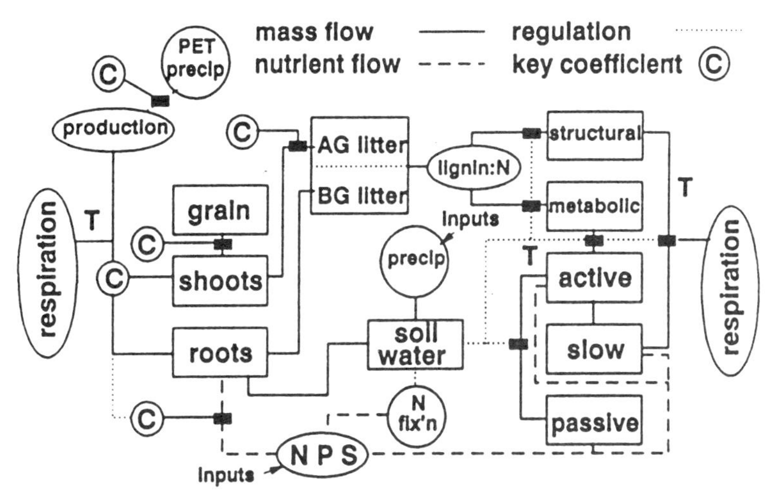

Figure L.2 An example of the linkage between crop productivity and soil organic matter C and N subroutines in CENTURY (after Woomer, 1992).

References

Burke, I.C., Schimel, D.S., Yonker, C.M., Parton, W.J., Joyce, L.A. and Lauenroth, W.K. (1990) Regional modelling of grassland biogeochemistry using GIS. *Landscape Ecology* **4**, 45-54.

Jenkinson, D.S. and Rayner, J.H. (1977) The turnover of soil organic matter in some of the Rothamsted classical experiments. *Soil Science* **123**, 298-305.

Parton, W.J., Schimel, D.S., Cole, C.V. and Ojima, D.S. (1987). Analysis of factors controlling soil organic matter levels in Great Plains grasslands. *Soil Science Society of America Journal* **51**, 1173-1179.

Parton, W.J., Sanford, R.L., Sanchez, P.A. and Stewart, J.W.B. (1989) Modelling Soil Organic Matter Dynamics in Tropical Soils. In: Coleman, D.C., Oades, J.M. and Uehara, G. (eds) *Dynamics of Soil Organic Matter in Tropical Ecosystems.* NifTAL Project, Paia, Hawaii. pp153-171.

Van Veen, J.A. and Paul, E.A. (1981) Organic Carbon Dynamics in grassland soils. 1. Background information and computer simulation. *Canadian Journal of Soil Science* **61**, 185-201.

Woomer, P.L. (1992) Modelling organic matter dynamics in tropical ecosystems: model adoption, uses and limitations. In: Molongoy, K. and Merckx, R., (eds.) *Organic Matter Dynamics in Tropical Ecosystems*, prodeedings of a symposium held at Katholieke University, Leuven, Belgium, 6-9 November, 1992. (In press).

Woomer, P.L. and Ingram, J.S.I. (1990) *The Biology and Fertility of Soils*: TSBF Report 1990. TSBF, Nairobi, Kenya.

Appendix M QUANTIFICATION OF NEAR-SURFACE MICROCLIMATE

Alan Stapleton

M.1 INTRODUCTION

The purpose of this appendix is to provide guidelines for the measurement of microclimate both above and below ground. These measurements are required for studies of management practices employed by farmers to preserve or enhance the beneficial effects of variations in microclimate and those employed to mitigate against adverse microclimatic factors. The rates of biological processes in the soil are controlled by microclimate rather than macroclimate.

Microclimate is influenced by local topography, vegetation and soil type. A distinct microclimate exists close to the soil surface, and amongst plants where vegetation cover influences light, temperature, humidity and wind speed compared to an open site. Below ground the soil also has a very distinct climate characterized by soil temperature, soil humidity or water content and aeration.

Microclimate influences farming practices and these influences merit investigation in the TSBF Programme as an understanding of them may provide options for improving crop productivity. Stigter *et al.* (1986) distinguish between microclimate modification and manipulation. Microclimate modifications are inadvertent changes in surface exposure resulting, for example, from weeding or burning. Microclimate manipulations are deliberate actions to alter severe aspects of the microclimate by for example mulching or shading.

M.2 MICROCLIMATIC FACTORS

M.2.1 Parameter selection

In microclimate investigation those factors expected to be agronomically decisive should be selected for quantification. As the importance may well be indirect, agronomic significance may have to be demonstrated through field experiments conducted over several seasons. In on-farm investigations comparison of different management practices requires random sampling of different microclimatic conditions. Sampling strategies for micrometeorological experiments are described by Stigter *et al.* (1976).

M.2.2 Meteorological observations

Microclimates are largely a consequence of the interaction of soil or plant surfaces with large scale meteorological conditions. The meteorological observations used for TSBF site characterisation are fully described in Section 2.2 in the main text. Direct observations consist of precipitation (rainfall), air temperature (measured 1.2 - 2.0 m above ground), relative air humidity, incidence of ground frost, solar radiation or sunshine, wind speed (measured 2 m above ground) and pan evaporation. From these observations potential evapotranspiration for grass as a reference crop or open water evaporation are calculated using Penman type combination equations based on energy balance considerations. Microclimatic observations are made using similar equipment. Further details of instruments particularly suitable for agrometeorological field stations is given by Doorenbos (1976).

M.2.3 Micrometeorological measurements

Over the last twenty years, Stigter, working in a range of tropical countries has made extensive reviews of the literature concerning microclimate modification and management. This work provides information about measurement methods appropriate for microclimate measurement in the study of the major management practices that affect microclimate: these are mulching, shading and wind protection.

M.2.4 Mulching

The main effects of mulching on microclimate are mediation of temperature in both the mulch and in the soil surface, and a reduction water loss by evaporation (Stigter, 1984b and 1988a). Temperature is measured using thermister probes or thermometers. If exposed at or above the ground surface, thermometers need a cover to protect them from solar radiation but must allow air circulation. Diurnal temperature fluctuation has been found to be a more sensitive indicator of mulch effects than average soil temperature (Othieno *et al.*, 1985). Soil water content can be measured gravimetricaly, or after appropriate calibration by neutron probe (Long and French, 1967; Greacen, 1981), nylon resistance units or gypsum blocks (Bouyoucos, 1954). Alternatively the water content of relatively moist soils can be determined indirectly from soil water potential measured using tensiometers (Richards, 1942). Continuous monitoring is only possible for gypsum blocks and pressure transducer tensiometers.

M.2.5 Shading

Shading has the following effects which have been reviewed by Stigter (1984a and 1988a): Protection from the sun, reduction of longwave radiation loss to the atmosphere, weed suppression, reduction of light and radiation in mixed crops and alley cropping. Shading may also have detrimental effects and pruning may be required to ensure that shaded crops receive adequate light.

The main factors that are altered under shade are temperature and radiation. Radiation may be measured using point sensors, moving sensors or sensors that integrate over an area. Solarimeters or pyranometers measure global solar radiation (W/m^2) in the 300-1100 nm waveband and are suitable for continuous monitoring. Stigter and Musabilha (1982) used a point sensor to record the near infra-red solar radiation and determined photosynthetically active radiation (PAR) by difference. Sensors are now available that measure PAR directly as radiation in the 400-700 nm waveband and quantum sensors are available that are sensitive to photons. The higher efficiency of red photons is taken into account thereby giving a more accurate estimate of the photosynthetic activity of radiation reaching the crop. Moving sensors have been used to measure radiation over a larger area (Stigter *et al.*, 1977) but these require an electricity supply for moving the sensors. Integrating sensors have been used by Newman (1985) to measure photosynthetically-active radiation.

Quantitative measurement of net longwave radiation at night has not been successful yet, unprotected minimum thermometers should give a more meaningful comparison in such cases.

M.2.6 Wind protection

The effects of shelter introduced for protection against mechanical damage to trees and crops, and to reduce water stress and wind erosion are reviewed by Stigter (1984c) and Stigter *et al.* (1988b). Although the spatial pattern of wind protection is well known in front of and behind shelterbelts, the effects of trees in forest edges, and the pattern of protection given by single or scattered trees has not been studied. Protection from wind is quantified by measurement of windspeed (m/s). Specially adapted cup anemometers need to be used as the type used for meteorological observations are not sufficiently accurate for comparing treatment effects. Shaded Piche evaporimeters (Stigter and Uiso, 1981) have been used as alternative simple air movement meters. Measurements were made about 20 cm above the protected surface.

M.3 PROCESSES AFFECTING MICROCLIMATE

M.3.1 Radiation balance

The energy which heats crop and soil surfaces and evaporates water from them comes from solar radiation and longwave radiation from the atmosphere. Longwave radiation is also emitted from the surface. Campbell (1985a) provides computational procedures for deriving equations for heat and vapour exchange at crop and soil surfaces using daily observations of solar radiation and air temperature. If solar radiation measurements are not available estimates can be made from observations of sunshine hours using standard constants given by Doorenbos and Pruitt or specific constants for different climatic regions (Frere and Popov, 1979).

M.3.2 Soil temperature and heat flow

All of the physical, chemical and biological processes that go on in the soil are influenced by soil temperature and in some cases growth of above-ground plant parts is more closely correlated with soil temperature than air temperature. Soil temperatures may be calculated as a function of depth and time by numerical solution of the heat flow equation. Campbell (1985b) provides values for calculating soil thermal conductivity and specific heat from proportions of mineral, water, air and organic matter in soils and a procedure for defining the boundary condition at the soil surface.

M.3.3 Soil aeration

Oxygen, carbon dioxide and water vapour diffuse into and out of the soil. In well aerated soil the carbon dioxide content is around 0.25% and the oxygen content is 20.73%. Oxygen contents below 10-15% can inhibit plant growth. Campbell (1985c) provides a computational scheme based on Fick's Law for estimating oxygen concentrations and fluxes in the soil as a function of aeration porosity and depth below the surface. Practices leading to soil compaction reduce aeration porosity and hence diffusion. A decrease in oxygen concentration is also the natural consequence of water-filled pores. Oxygen concentrations decrease towards the anaerobic centres of microbial and soil material when pores within them are water-filled.

M.3.4 Crop evapotranspiration

Methods for estimating evapotranspiration from meteorological observations have been developed by irrigation engineers and agronomists for determining crop water requirements. The guidelines published by the FAO (Doorenbos and Pruitt, 1977) have now been in use in many countries for some time. Where daily observations of mean air temperature, relative humidity (or vapour pressure), sunshine hours (or solar radiation) and windspeed are available the Penman combination equation is used to calculate potential evapotranspiration for grass as a reference crop. Crop factors which vary throughout the growing season are applied to estimate potential crop evapotranspiration. Calculation can be on a daily or weekly basis. Wherever possible locally derived crop factors should be used, but these are rarely available and factors provided in the FAO Guidelines are used for determining crop water requirements.

More detailed investigations of crop water use especially under water-limiting conditions require separation of evaporation from plant transpiration. Even when soil is covered by vegetation evaporation is probably at least 10% of evapotranspiration. A complete analysis of evaporation (Campbell, 1985d) takes into account both liquid and vapour flow and incorporates the effects of temperature gradients on water movement. This analysis is more complicated than required in many applications. If a water balance rather than water content profiles are required then vapour flow except at the surface might be neglected.

M.4 LINKED TRANSPORT MODELS

In a wide variety of field and laboratory investigations concerned with the interdepence of physical, biological and chemical processes simulation models have helped to guide our thinking and experimental design. Pioneering work on modeling evapotranspiration, soil water and solute flux and plant uptake was conducted at Utah State University (Hanks and Bowers, 1962; Nimah and Hanks, 1973). More recently work at the Swedish University of Agricultural Sciences has led to the development of SOIL, a linked transport model that simulates the flow of both heat and water in soils (Jansson, 1991). This model is particularly appropriate for using in the investigation of management factors that influence microclimate. The driving variables needed to simulate both soil heat and water flows in a natural vegetated soil are precipitation, air temperature, relative humidity, wind speed and solar radiation. A number of physical soil parameters are required; a good soil profile description combined with the soil physical characterization given in Chapter 7 will provide the information required.

Measurements of leaf area index (or plant cover, surface resistance and interception are required as these are important controls of water loss to the atmosphere and depth of litter and humus are required as these are important controls of soil heat balance.

M.5 AUTOMATIC RECORDING OF MICROCLIMATIC FACTORS

Instrumentation for continuous measurement of micrometeorological factors is continually under development. Two basic types of recording instruments are now available. These are data loggers to which a variety of sensors can be connected and intelligent sensors which have storage facility built in and capacity for connecting additional sensors.

Loggers are available with up to 32 analogue channels and 3 digital channels. For micrometeorological investigation sensors such as temperature probes, combined temperature and relative humidity probes, tipping bucket raingauges, cup anenometers, pyranometers, soil temperature probes, soil moisture blocks and tensiometers might be employed. To calculate the storage capacity required of a logger, multiply the number of sensors by the readings required per day by the number of days the logger will be left unattended. A 32000 record storage capacity will be adequate for half hourly readings of up to 30 sensors stored over two weeks. Data are analysed on an office based computer using software provided with the logger. Inexpensive portable computers or personal organisers can be used to transfer data from the logger to the office based computer using a cable connected to the RS232 port on the computer.

Intelligent sensors commercially available include tipping bucket raingauges, combined temperature and relative humidity meters and pyranometers. Each intelligent unit can support up to three additional sensors such as temperature probes, tensiometers or additional light sensors. Alternatively a four channel intelligent unit can support up to four temperature probes or tensiometers. An advantage of these sensors is the portability of the unit which can be connected to the RS232 port of a computer in the office. Alternatively a portable computer could be taken to the field to collect the stored data.

For any experimental situation the most suitable recording system will be influenced by the availability of meteorological observations nearby and the computer facilities available. For microclimatic studies under a forest canopy near The University of Exeter we have selected a recording system that consists of an intelligent sensor to measure rainfall and air temperature to augment local meteorological observations and also a data logger for recording air and soil temperatures and soil water content under the canopy.

References

Bouyoucos, G.J. (1954) Electrical Resistance Methods as finally perfected for making continuous measurement of soil moisture content under field conditions. *Michigan State College Quarterly Bulletin* **37**, 132-149.

Campbell, G.S. (1985a) Atmospheric Boundary Conditions. In: *Soil Physics with BASIC.* Transport Models for Soil-Plant Systems. Elsevier. pp.134-145.

Campbell, G.S. (1985b) Soil Temperature and Heat Flow. In: *Soil Physics with BASIC.* Transport Models for Soil-Plant Systems. Elsevier. pp.26-39.

Campbell, G.S. (1985c) Gas Diffusion in Soil. In: *Soil Physics with BASIC.* Transport Models for Soil-Plant Systems. Elsevier. pp.12-25.

Campbell, G.S. (1985d) Evaporation. In: *Soil Physics with BASIC.* Transport Models for Soil-Plant Systems. Elsevier. pp.98-107.

Doorenbos, J. (1976) *Agro-meteorological field stations.* FAO Irrigation and Drainage Paper 27, FAO, Rome.

Doorenbos, J. and Pruitt, W.O. (1977 revised) *Guidelines for Predictoing Crop Water Requirements.* FAO Irrigation and Drainage Paper 24. Food and Agriculture Organisation, Rome.

Frere, M. and Popov, G. (1979) *Agrometeorological Crop Monitoring and Forecasting.* FAO Plant Production and Protection Paper No 17. Food and Agriculture Organisation, Rome.

Greacen, E.L. (ed.) (1981) *Soil Water Assessment by the Neutron Method.* CSIRO, Australia, 149pp.

Hanks, R.J. and Bowers, S.A (1962) Numerical solution of the moisture flow equation for infiltration into layered soils. *Soil Science Society of America Proceedings* **26**, 530-534.

Jansson, P-E. (1991) *Simulation Model for Soil Water and Heat Conditions - Description of the SOIL Model.* Swedish University of Agricultural Sciences. Uppsala.

Long, I.F. and French, B.K. (1967) Measurement of soil moisture in the field by neutron moderation. *Journal of Soil Science* **18**, 149-166.

Newman, S.M. (1985) Low cost sensor integrators for measuring transmissivity of complex canopies to photosynthetically active radiation. *Agriculture, Forestry and Meteorology* **35**, 243-254.

Nimah, M.N. and Hanks, R.J. (1973) Model for estimating soil water, plant and atmospheric interrelations: 1 Description and sensitivity. *Soil Science Society of America Proceedings* **37**, 522-527.

Othieno, C.O, Stigter, C.J. and Mwampaja, A.R. (1985) On the use of Stigter's ratio in expressing thermal efficiency of grass mulches. *Experimental Agriculture* **21**, 169-174.

Richards, L.A. (1942) Soil moisture tensiometers: Materials and construction. *Soil Science* **53**, 241-248.

Stigter, C.J. (1984a) Traditional Use of Shade: A method of microclimate modification. *Archives of Meteorology, Geophysics and Bioclimatology* **34** 203-210.

Stigter, C.J. (1984b) Mulching as a traditional method of microclimate management. *Archives of Meteorology, Geophysics and Bioclimatology* **35** 147-154.

Stigter, C.J. (1984c) Wind protection in traditional microclimate management and manipulation: examples from East Africa. In: Grace, J. (ed.), *The Effects of Shelter on the Physiology of Plants and Animals. Progress in Biometeorology* **2**, 145-154, Swetz and Zeitlinger, Lisse, Netherlands.

Stigter, C.J. (1988) *Microclimate management and manipulation in agroforestry: Viewpoints on Agroforestry.* In: Wiersum, K.F. (ed.). Agricultural University, Wageningen, Netherlands.

Stigter, C.J. and Musabilha, V.M.M. (1982) The conservative ratio of photosynthetically active to total radiation in the tropics. *Journal of Applied Ecology* **19**, 853-858.

Stigter, C.J. and Uiso, C.B.S. (1981) Understanding the Piche evaporimeter as a simple integrating mass transfer meter. *Applied Scientific Research* **37**, 213-224.

Stigter, C.J. Lengkeek, J.G., and Kooijman, J. (1976) A simple worst case analysis for estimation of correct scanning rate in a micrometeorological experiment. *Netherlands Journal of Agricultural Science* **24**, 3-16.

Stigter, C.J., Goudriaan, J. Bottemanne, F.A., Birnie, J., Lengkeek, J.G. and Sibma, L. (1977) Experimental evaluation of a crop climate simulation model for Indian corn (*Zea mays* L.). *Agricultural Meteorology* **18**, 163-166.

Stigter, C.J., Darnhofer, T., Karing, P.H., Lawson, T.D., Popov, G.F. and von Hoy ningen-Huene, J. (1986) *Microclimate management and manipulation in Traditional Farming.* Report of the Working Group of microclimate management and manipulation in traditional farming. Ninth Session of Commission for Agric. Meteorology, Madrid. WMO, Geneva. 74pp.

Stigter, C.J., Darnhofer,T. and Herrera, S.H. (1988) Crop protection from very strong winds: recommendations in a Costa Rican agroforestry case study. In: Reifsnyder, W.E. and Darnhofer, T. (eds), *Application of Meteorology in Agroforestry Systems Planning and Management.* Proceedings of an ICRAF/WMO/UNEP Workshop, Nairobi, Kenya.

Appendix N SUGGESTED EQUIPMENT AND REAGENT LISTS FOR ROUTINE TSBF ANALYSES

John Ingram

N.1 EQUIPMENT AND NON-REAGENT CONSUMABLES

This list is intended to serve as a basis for those wishing to establish a laboratory (or upgrade obsolete equipment) so as to undertake measurements for TSBF studies. Prices (see below) are given to assist with budgeting. Although lengthy, the list is not exhaustive.

The quantity (q) given for each item is a guide, but may be considered to be the minimum for a lab analysing perhaps 500 - 1000 samples per year, in batches of 24 (including 2 repeats and 2 blanks), for most of the analyses in this Handbook. The unit prices (up; UK£) are correct for April 1992 for the quoted suppliers (s; see Section N.2) but are subject to change. The indicated items are by no means the only product that can be used for a particular task, but they have been found to be very satisfactory; similar (or identical) items can be obtained from various other suppliers both in the UK and elsewhere, but of course the prices may vary. It is strongly recommended that an updated/recent quotation is obtained prior to placing orders.

Note: When ordering electrical equipment remember to specify both the voltage (V) and A.C. current frequency (Hz) required.

i Item
s Supplier
c Supplier's code
q Quantity
up Unit price (UK£)
tp Total line price (UK£)

Note: The price (in UK £) does not include packing, freight or insurance.

	i	s	c	q	up	tp
1.	AAS Igniter	U	10171	1	12.00	12.00
2.	AAS Air filter	U	BS303-0229	1	85.00	85.00
3.	AAS Air compressor	U	BS303-0313	1	462.00	462.00
4.	AAS Air filter (replacement)	U	BS303-2471	1	7.00	7.00

5.	AAS Acetylene regulator	U	BS303-0106	1	100.00	100.00
6.	AAS Exhaust system	U	BS303-1406	1	212.00	212.00
7.	AAS (with Ca/Mg lamp) (remember to specify gas fitting sizes)	U	Buck 200A	1	6545.00	6,545.00
8.	Aspirator, polythene, 9 litre	F	ASP-510-030W	5	8.58	42.90
9.	Aspirator, polythene, 23 litre	F	ASP-560-010Q	5	10.36	51.80
10.	Balance 200 x 0.01g	F	BCR-320-A	2	525.30	1,050.60
11.	Balance 600 x 0.1g	F	BCR-330-Q	1	364.62	364.62
12.	Balance 310 x 0.001g	F	BFH-650-010T	1	1310.78	1,310.78
13.	Beaker, Pyrex, 100ml (10 pack)	F	BNB-300-090M	5	17.55	87.75
14.	Beaker, Pyrex, 600ml (10 pack)	F	BNB-320-190Y	1	28.55	28.55
15.	Beaker, Pyrex, 1000ml (5 pack)	F	BNB-300-230W	2	23.93	47.86
16.	Beaker, plastic, 5 litre	F	BNH-700-170F	5	14.47	72.35
17.	Bottle, screw cap (30ml) (for ground soil storage) (40 pack)	F	BTF-600-030G	5	10.63	53.15
18.	Bottle, wide mouth, 500ml (for texture analysis) (20 pack)	F	BTF-600-130C	3	12.32	36.96
19.	Brush, test tube, nylon, 30mm (10 pack)	F	BUR-580-030L	5	2.73	13.65
20.	Brush, flat form, bristle	F	BUR-750-C	5	1.28	6.40
21.	Burette, 5ml x 0.05ml	F	BWF-605-050G	5	10.73	53.65
22.	Burette, 25ml x 0.1ml	F	BWF-610-110X	5	8.65	43.25
23.	Burette, 50ml x 0.1ml	F	BWF-610-150C	5	9.06	45.30
24.	Centrifuge bucket, (6 pack)	B	306/0071/36	1	87.00	87.00
25.	Centrifuge (Denly MkIV)	B	306/0071/00	1	1226.00	1,226.00
26.	Centrifuge angle head (6x50ml)	B	306/0071/11	1	186.00	186.00
27.	Centrifuge tubes (25 pack) 50ml	B	402/0325/06	4	45.00	180.00

28.	Crucible lids (10 pack)	F	CWB-720-531D	3	12.74	38.22
29.	Crucible (10 pack)	F	CWB-710-030J	3	11.84	35.52
30.	Cylinder, measuring, 10ml (3 pack)	F	CYL-302-030G	2	4.12	8.24
31.	Cylinder, measuring, 100ml (2 pack)	F	CYL-302-090L	5	4.53	22.65
32.	Cylinder, measuring, 1000ml (2 pack)	F	CYL-302-150T	5	16.48	82.40
33.	Deioniser	F	DCF-500-010W	1	356.25	356.25
34.	Deioniser, resin refill (2 pack)	F	DCF-505-010P	2	96.90	193.80
35.	Deioniser, spare batteries	F	BMT-540-050J	2	0.91	1.82
36.	Desiccator, glass	F	DES-250-050Y	2	68.55	137.10
37.	Diluter, Brand	F	DHL-250-X	2	238.78	477.56
38.	Dispenser, 2-10ml, Brand	F	DHT-350-050D	2	121.58	243.16
39.	Dispenser, 5-25ml, Brand	F	DHT-350-070U	2	168.99	337.98
40.	Dispenser, 10-50ml, Brand	F	DHT-350-090X	2	235.10	470.20
41.	Dispenser bottle, 1000ml	F	BTF-680-130P	10	4.66	46.60
42.	Dispenser bottle, 2000ml	F	BTF-680-150J	5	10.79	53.95
43.	Distilled water plant	F	WGS-522-010Y	1	448.57	448.57
44.	Distilled w.p., spare boiler	F	W400/B	1	130.78	130.78
45.	Distilled w.p., spare condenser	F	WC48/M2	1	155.19	155.19
46.	Distilled w.p., spare heater + thermostat	F	A6/6	1	132.10	132.10
47.	Drying cabinet, 120°C	F	OVL-235-010Y	1	1311.00	1,311.00
48.	Drying cabinet, extra shelf	F	OVL-240-071L	1	46.00	46.00
49.	Drying tray 200 x 300 x 50cm (galvanised metal or wood)	L		100	1	100.00

50.	EC meter	F	CRT-500-010F	1	266.46	266.46
51.	EC meter, temperature probe	F	CRT-505-500E	1	54.90	54.90
52.	EC meter, battery	F	BMT-680-090H	2	2.81	5.62
53.	Flame photometer	F	FGA-351-F	1	2277.33	2,277.33
54.	Flame photometer compressor	F	PXW-790-W	1	179.64	179.64
55.	Flask, Erlenmeyer, Pyrex, 100ml (10 pack)	F	FHB-375-090L	6	17.92	107.52
56.	Flask, Erlenmeyer, Pyrex, 250ml (10 pack)	F	FHB-375-130C	3	19.41	58.23
57.	Flask, Erlenmeyer, Pyrex, 500ml (10 pack)	F	FHB-375-150T	1	26.82	26.82
58.	Flask, Buchner, Pyrex, 500ml	F	FHD-355-070L	3	20.46	61.38
59.	Flask, volumetric, 100ml (5 pack)	F	FHM-365-150Y	6	29.17	175.02
60.	Flask, volumetric, 250ml (2 pack)	F	FHM-365-190T	15	16.19	242.85
61.	Flask, volumetric, 1000ml (2 pack)	F	FHM-365-230K	4	28.72	114.88
62.	Fritch P15 Plant grinder (with collector and 0.25mm sieve)	M	Pulverisette15	1	3250.00	3,250.00
63.	Fritch stand for grinder	M	Stand	1	353.00	353.00
64.	Funnel (3"), plastic, (10 pack)	F	FPH-450-030J	5	2.89	14.45
65.	Funnel, Buchner, 90mm	F	FPL-390-090A	3	17.83	53.49
66.	Funnel, separating, 250ml	F	FPM-620-090P	2	20.12	40.24
67.	Heater block 100-450°C (with insert for 12 tubes, 12 tubes and rack)	T	DB4/26	2	2320.00	4,640.00
68.	Heater block, spare tubes (6 pack)	T	F6570	4	139.00	556.00
69.	Hydrometer, jar (60 x 440 mm)	F	HYD-800-130R	25	15.38	384.50
70.	Hydrometer, soil	F	SLR-250-G	2	31.43	62.86

71.	Magnetic stirrer/heater	F	SWT-515-010U	2	239.20	478.40
72.	Magnetic follower 45mm (3 pack)	F	SWX-260-130M	2	2.79	5.58
73.	Magnetic follower retriever	F	SWX-510-N	2	5.30	10.60
74.	Mains plug 13 A	F	ECA-352-J	10	0.78	7.80
75.	Micropipette, Brand (0.1ml)	F	PMP-721-110T	2	97.41	194.82
76.	Micropipette, tips (1000 pack)	F	PMP-724-502H	2	11.30	22.60
77.	Micropipette, 1-5ml	F	PMP-850-050L	1	136.99	136.99
78.	Micropipette, tips (10 pack)	F	PMP-772-506W	5	10.66	53.30
79.	Microscope, dissecting/light	F	MIB-ABA-020M	1	178.10	178.10
80.	Mixing plunger, sedimentation	L		1	3	3.00
81.	Muffle furnace 13A, 2.5 litre	F	FSE-285-010A	1	1088.00	1,088.00
82.	Oven, drying/fan, 97 litre	F	OVB-306-110T	1	847.36	847.36
83.	Pestle and mortar	F	MWA-300-071S	3	11.03	33.09
84.	pH meter, bench	F	PHL-203-010W	1	683.20	683.20
85.	pH meter, portable	F	PHK-080-010K	2	44.65	89.30
86.	pH meter, portable, battery (4 pk)	F	BMT-620-120U	1	3.59	3.59
87.	pH electrode	F	PHQ-300-030D	3	72.46	217.38
88.	Photocopier (e.g. Canon NP 112)	L		1	1000.00	1,000.00
89.	Pipette, 1ml (2 pack)	F	PMC-285-020J	5	12.56	62.80
90.	Pipette, 2ml (2 pack)	F	PMC-285-030G	5	12.56	62.80
91.	Pipette, 5ml (2 pack)	F	PMC-285-070R	5	13.55	67.75
92.	Pipette, 10ml (2 pack)	F	PMC-285-080X	5	14.91	74.55
93.	Pipette, 25ml (2 pack)	F	PMC-285-110Y	5	18.34	91.70
94.	Pipette filler, rubber	F	PMR-330-090E	3	5.93	17.79

95.	Plastic buckets	L		5	3	15.00
96.	Voltage surge protectors	F	ECA-780-X	5	16.24	81.20
97.	Polythene bottle, 60ml (10 pack)	F	BTK-390-050D	10	3.51	35.10
98.	Polythene bottle, 150ml (10 pack)	F	BTK-390-070U	20	4.40	88.00
99.	Refrigerator, 215 litre	F	RFP-300-010N	1	367.71	367.71
100.	Safety lab coats	L		10	5	50.00
101.	Safety eye wash station	F	SAP-740-W	1	16.83	16.83
102.	Safety eye wash bottle	F	SAP-714-T	2	5.23	10.46
103.	Safety BCF fire extinguisher	F	SAP-870-Q	1	64.78	64.78
104.	Safety eye specs.	F	SAP-290-E	10	1.80	18.00
105.	Safety gloves, PVC	F	SAR-890-070H	2	5.40	10.80
106.	Safety gloves, latex (50 pack)	F	SAR-710-051G	1	28.71	28.71
107.	Shaker, SM25	F	SGM-540-025T	1	1952.23	1,952.23
108.	Sieve, brass, 0.25mm	F	SIH-400-220G	2	36.92	73.84
109.	Sieve, brass, lid	F	SIH-470-504M	2	16.73	33.46
110.	Sieve, S.Steel, 2mm	F	SIH-460-100U	2	44.41	88.82
111.	Sieve, receiver	F	SIH-470-506Q	2	19.47	38.94
112.	Spectrophotometer, Coil	C	CE 2010	1	2640.00	2,640.00
113.	Spectrophotometer Micro Sipette control	C	2020 2100	1	1410.00	1,410.00
114.	Spectrophotometer Holder for Micro Sipette cell	C	1010 3200	1	165.00	165.00
115.	Spectrophotometer Micro Sipette Flow cell (Silica 20mm)	C	203 0703	1	390.00	390.00
116.	Spectrophotometer, 10mm cell (for use without Micro Sipette system)	C	303 0726	3	14.00	42.00

117.	Spectrophotometer, spare fuses	C	2020 2700	1	23.00	23.00
118.	Spectrophotometer, spare lamp (2 pack)	C	2303 0140	1	75.00	75.00
119.	Spin mixer	F	SGP-202-010J	2	173.04	346.08
120.	Stoppers (50 pack)	F	BTF-376-090T	1	5.69	5.69
121.	Stopwatch	F	TKP-386-G	1	11.25	11.25
122.	Test tube 150 x 16mm (100 pack)	F	TES-200-191R	5	9.81	49.05
123.	Test tube rack, 16mm test tube (24 place)	F	STK-640-070L	10	4.85	48.50
124.	Test tube holder (10 pack)	F	TEW-710-J	1	7.79	7.79
125.	Thermometer -10 - 50°C (10 pk) (for sedimentation)	F	THL-210-031A	1	15.57	15.57
126.	Thermometer -10 - 400°C (10 pk) (for block heaters)	F	THL-210-131T	1	15.14	15.14
127.	Tongs	F	TNS-200-010R	5	2.78	13.90
128.	Tray, polystyrene, (3 pack)	F	TRC-200-090R	10	11.59	115.90
129.	Trolley	F	TSF-351-A	1	134.31	134.31
130.	Tubing, PVC, 10m x 8 x 2mm	F	TWR-670-170J	3	2.84	8.52
131.	Tubing connectors, T, 9 mm	F	ADF-715-070W	1	8.65	8.65
132.	UPS Emerson 500VA	E	Model 30	1	321.00	321.00
133.	Vacuum cleaner, cylinder	L		1	100	100.00
134.	Vacuum pump, oil free	F	PYB-530-010A	1	1001.73	1,001.73
135.	Vacuum pump, spare valve seal	F	PYB-536-020H	2	19.12	38.24
136.	Vacuum pump, spare diaphragm	F	PYB-536-010K	2	23.90	47.80
137.	Vacuum pump, water, filter	F	PXY-311-P	2	7.62	15.24
138.	Vacuum pump, water, connectors	F	PXY-315-500R	2	1.31	2.62

139.	Washbottle, 250ml	F	WBS-400-020C	10	0.72	7.20
140.	Washbottle, 500ml	F	WBS-400-040T	10	0.86	8.60
141.	Water tank (450 x 1400 x 400 cm deep)	L		1	10	10.00
142.	Water bath, 22 litre	F	BJL-455-210F	1	450.85	450.85
143.	Cotton wool, absorbent, 0.5kg	F	CTC-270-Q	4	2.64	10.56
144.	Filter paper, Whatman 1 9.0cm	F	FHD-240-110F	10	2.14	21.40
145.	Filter paper, Whatman 1 15.0cm	F	FHD-240-170K	10	3.71	37.10
146.	Filter paper, Whatman 42 9.0cm	F	FHD-460-110V	10	6.59	65.90
147.	Filter paper, Whatman 42 15.0cm	F	FHD-460-170D	10	15.26	152.60
148.	Pen, marker, red, (12 pack)	F	LAC-710-020V	1	12.96	12.96
149.	Pen, marker, blue, (12 pack)	F	LAC-710-030S	1	12.96	12.96
150.	Pen, marker, black, (12 pack)	F	LAC-710-050M	1	12.96	12.96

N.2 SUGGESTED EQUIPMENT SUPPLIERS

B BDH Laboratory Supplies Apparatus Division, Merck House, Poole, BH15 9EL, UK, TEL +44 202 669700, TLX 264772 bdh app, FAX +44 202 666542

C Cecil Instruments Limited, Milton Technical Centre, Cambridge Road, CB4 6AZ, UK, TEL +44 223 420821, FAX +44 223 420475

E Computer Battery Services Ltd, Unit 8, Riverside Park, Farnham, Surrey GU8 7UG, UK, Tel +44 252 714100, Fax +44 252 733910

F Fisons Ltd, Bishop Meadow Road, Loughborough, Leics LE11 0RG, UK, TEL +44 509 231166, FAX +44 509 231893

L Local

M Christison Scientific Equipment Ltd, Albany Road, Gateshead, NE8 3AT, UK, TEL +44 91 477 4261, FAX +44 91 490 0549

U Buck Scientific, Unit 6, Upper Wingbury Courtyard, Upper Wingbury Farm, Wingrave, Aylsbury, Bucks, UK, TEL +44 402 349136, FAX +44 296 681293

N.3 REAGENTS

This list gives the minimum quantities (or the smallest commercially available quantity, whichever is less) of the major reagents required to analyse 100 samples for each method in this Handbook. The unit quoted prices (up; UK£) are correct for April 1992, from the quoted supplier (Merck UK) but are subject to change; alternative reputable suppliers are however equally good, and prices may vary. It is strongly recommended that a recent quotation is obtained prior to placing orders.

s Supplier
c Supplier's code
q Quantity
up Unit price (UK£)
tp Total line price (UK£)

	s	c	q	up	tp
Acetic acid, Analar	500ml	100013L	1	6.10	6.10
Acetone	1000ml	270235S	2	5.40	10.80
Ammonium acetate	500g	271424C	10	6.80	68.00
Ammonium chloride, Analar	500g	100173D	1	6.60	6.60
Ammonium molybdate	100g	271872W	1	8.70	8.70
Ammonium sulphate, Analar	500g	100333B	1	6.60	6.60
Amyl alcohol	500ml	272124U	1	4.70	4.70
Ascorbic acid	100g	103033E	1	7.60	7.60
Barium chloride	500g	272904R	1	6.60	6.60
Boric acid	500g	274104B	1	4.30	4.30
Buffer tablets (pH 4)	50 pack	331542Q	1	10.90	10.90
Buffer tablets (pH 7)	50 pack	331552S	1	10.50	10.50
Buffer tablets (pH 9.2)	50 pack	331562U	1	8.50	8.50
Calcium carbonate, Analar	250g	100683U	1	7.80	7.80
Cetyltrimethylammonium bromide	500g	276654L	1	23.90	23.90
Chloroform	1000ml	277105X	1	9.70	9.70
Devarda's alloy	500g	280094F	1	25.10	25.10
Ethanol*	2500ml	283047K	1	16.00	16.00
Ethanol	500ml	283047K	1	16.00	16.00
Ferric nitrate	500g	283844Y	1	13.10	13.10
Ferrous ammonium sulphate	500g	271664Q	1	5.90	5.90
Hydrochloric acid, Analar*	2500ml	103076P	1	8.50	8.50
Hydrochloric acid, Analar	1000ml	101255Y	3	5.90	17.70
Hydrogen peroxide (30%)¶	1000ml	286945B	1	7.20	7.20
Lanthanum chloride	100g	103433Q	1	19.80	19.80
Lithium sulphate, Analar	500g	101474R	1	18.30	18.30
Magnesium sulphate, Analar	500g	101514Y	1	7.00	7.00
Methanol*	2500ml	291926G	1	8.60	8.60
Methanol	500ml	291924E	5	3.50	17.50
Methoxyethanol	500ml	103624V	1	5.50	5.50
Methylpropan-2-ol	500ml	103584H	1	6.80	6.80

Octan-2-ol	2500ml	294106N	1	16.20	16.20
Orthophosphoric acid, Analar	500ml	101734S	1	6.90	6.90
Oxalic acid	500g	294234U	1	10.10	10.10
pH 4.5 indicator solution	500ml	210414M	1	8.10	8.10
Phenolphthalein	100g	200892K	1	7.00	7.00
Phosphomolybdic acid	100g	292713W	1	20.20	20.20
Potassium acetate	500g	295814P	1	5.10	5.10
Potassium chloride	1000g	295945C	5	5.10	25.50
Potassium chloride, Analar	250g	101983K	1	4.40	4.40
Potassium dichromate	500g	296054D	1	14.70	14.70
Potassium dihydrogen phosphate	500g	102034B	1	8.40	8.40
Potassium nitrate, Analar	250g	102143F	1	4.80	4.80
Potassium permanganate	500g	296444N	1	6.10	6.10
Potassium sulphate	2500g	296585C	2	11.20	22.40
Salicylic acid	500g	300384B	1	7.30	7.30
Selenium powder	100g	300454V	1	20.20	20.20
Silver nitrate	25g	300872M	1	24.30	24.30
Silver sulphate	25g	102343L	1	30.80	30.80
Sodium bicarbonate	500g	102474V	1	4.50	4.50
Sodium carbonate, Analar	500g	102394W	1	5.20	5.20
Sodium chloride, Analar	500g	102414J	1	4.50	4.50
Sodium citrate	500g	301284C	1	5.20	5.20
Sodium hexametaphosphate	500g	307344V	1	6.20	6.20
Sodium hydroxide	500g	301674M	1	4.50	4.50
Sodium hypochlorite solution	250ml	230393L	5	10.50	52.50
Sodium nitroprusside	100g	301902F	1	10.40	10.40
Sodium salicylate	500g	302104K	1	8.20	8.20
Sodium tartrate	500g	302274G	1	10.60	10.60
Sodium tungstate	500g	302374H	1	27.50	27.50
Sodium thiosulphate	1000g	302355E	1	4.60	4.60
Sucrose, Analar	500g	102744B	1	5.40	5.40
Sulphuric acid*	2500ml	303253D	4	7.91	31.64
Sulphuric acid	1000ml	303254E	10	5.44	54.40
Sulphuric acid, Analar	1000ml	102760B	2	7.40	14.80
Tannic acid	500g	303374L	1	19.80	19.80

Note: The price (UK£) does not include packing, freight or insurance.

* The 2500ml size is only allowed on a cargo aircraft; if this item is required to a destination where only a passenger service operates, select the smaller size given as the following item on the list.

¶ This product is not allowed on an aircraft; try to obtain locally.

N.4 SUGGESTED REAGENT SUPPLIER

BDH Laboratory Supplies, Merck Ltd., Poole, Dorset, BH15 1TD, UK; Tel +44 202 660444; Fax +44 202 666856; Tlx 41186 or 418123 tetra g.

Appendix O TSBF DATA SHARING POLICY

It is intended that participation in the TSBF Programme will require collaboration not only in methodology but also in the exchange of data. A clear statement about the ownership of data was deemed necessary.

1. All raw material and derived local data remain the property of the research worker who obtained it. No attempt to use such data will be made in TSBF publications, other than with the permission of the research worker concerned.

2. Each worker is encouraged to publish their work in the normal way, and to include reference to TSBF if they so wish. Reprints of, or references to, such works would be gratefully received by the TSBF Office, as part of the coordination exercise. TSBF would also aim to use such references to support applications to funding agencies; the more field studies that can be cited, the greater the chance of being awarded funds.

3. One of the aims of the TSBF Programme is to compare experimental variables measured in different ecosystems, and then to conduct syntheses of the data from each research site. Once synthesised, data from specific sites will lose their individuality, and become part of the TSBF regional, or ultimately global picture. Those conducting a synthesis exercise on behalf of TSBF may feel they have derived sufficient data to merit publication in its own right, and they should be encouraged to publish in an appropriate journal. Any such paper would include suitable acknowledgement of contributing research workers, and state that it constitutes part of the TSBF Programme. Should published data be used in the synthesis, references must be made in the normal style of a review. The authors will be encouraged to circulate a draft to those workers whose data were used in the synthesis, before submission to journals is made.